Basic Communication Skills
for Electronics Technology

Basic Communication Skills for Electronics Technology

ANDREA J. RUTHERFOORD
DeVry Institute of Technology
Atlanta, Georgia

Cartoons by Tom Ferguson
Drawings by Ty Johnson

PRENTICE HALL, Englewood Cliffs, New Jersey 07632

Library of Congress Cataloging-in-Publication and Data
Rutherfoord, Andrea J.
    Basic communication skills for electronics technology.

    Includes index.
    1. Technical writing. 2. Electronics—Authorship.
I. Title.
T11.R83   1989   808'.066621021   88-15122
ISBN 0-13-970617-8

Editorial/production supervision: *Mary Carnis*
Cover design: *Joel Mitnick Design, Inc.*
Manufacturing buyer: *Robert Anderson*

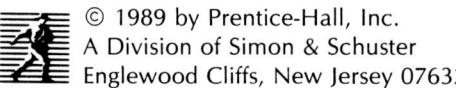

© 1989 by Prentice-Hall, Inc.
A Division of Simon & Schuster
Englewood Cliffs, New Jersey 07632

All rights reserved. No part of this book may be
reproduced, in any form or by any means,
without permission in writing from the publisher.

Printed in the United States of America

10 9 8 7 6 5 4 3 2 1

ISBN 0-13-970617-8

Prentice-Hall International (UK) Limited, *London*
Prentice-Hall of Australia Pty. Limited, *Sydney*
Prentice-Hall Canada Inc., *Toronto*
Prentice-Hall Hispanoamericana, S.A., *Mexico*
Prentice-Hall of India Private Limited, *New Delhi*
Prentice-Hall of Japan, Inc., *Tokyo*
Prentice-Hall of Southeast Asia Pte. Ltd., *Singapore*
Editora Prentice-Hall do Brasil, Ltda., *Rio de Janeiro*

CONTENTS

PREFACE TO THE TEACHER	xi
PREFACE TO THE STUDENT	xiii
ACKNOWLEDGMENTS	xv
MESSAGES FROM INDUSTRY	xvii

PART I
Foundations   1

CHAPTER 1 **GETTING STARTED**   3
Reading: *Technical Vocabulary* 3
Writing: Getting Started 7
Spelling: Plurals 13
Vocabulary: Latin and Greek Number Roots 16
Word Watch: A/An/And To/Two/Too 18

CHAPTER 2 **COMPLETENESS**   20
Reading: *Wardrobe: The First Step to Your Professional Image* 20
Writing: Topic Sentences 23
Spelling: Adding LY and LLY 25
Vocabulary: Negative Prefixes 26
Word Watch: Wear/Were/We're/Where 28

vi   *Contents*

CHAPTER 3	**CONCISENESS**	**29**
	Reading: *Shedding Light on Today's Lasers* 29	
	Writing: Writing a Summary 32	
	Spelling: IE/EI 34	
	Vocabulary: Roots SPEC and SON 35	
	Word Watch: They're/Their/There 35	
CHAPTER 4	**CLARITY**	**37**
	Reading: *Taking the Noise Out of Technical Writing* 37	
	Writing: Slang, Cliches, Analogies 41	
	Spelling: Doubling the Final Consonant 47	
	Vocabulary: SUB/SUPER Prefixes 48	
	Word Watch: Used, Supposed 49	
CHAPTER 5	**REVIEW**	**51**
	Reading: *Stress: A Deadly Wear and Tear* 51	

PART II
Planning a Technical Report   59

CHAPTER 6	**COMPARISON AND CONTRAST**	**61**
	Reading: *Grace, Style, and Intuition Contribute To Employee Success in Business* 61	
	Writing: Description, Comparison and Contrast 63	
	Spelling: Using Numbers 67	
	Vocabulary: Retro/Circum/Intro/Intra/Inter 70	
	Word Watch: T"OUGH" Words 71	
CHAPTER 7	**CAUSE AND EFFECT**	**73**
	Reading: *Introduction to the Scientific Method* 73	
	Writing: Cause and Effect/Formal Lab Reports 75	
	Spelling: Dropping the Final E/Slippery Silent Letters 84	
	Vocabulary: Other Number Prefixes (Mono/Bi/Semi/Poly) 85	
	Word Watch: Effect, Affect 86	
CHAPTER 8	**REVIEW**	**88**
	Reading: *Coping with Stress on the Job* 88	

Contents   vii

PART III
Writing a Technical Report   97

CHAPTER 9 **DESCRIPTIVE REPORTS**   **99**

Reading: *Working with Robots: The Real Story* 99
Writing: The Descriptive Report 102
Spelling: ANCE/ENCE Endings 107
Vocabulary: Proto/Trans/Neo 108
Word Watch: Accept/Except 109

CHAPTER 10 **PREPARING GRAPHICS**   **111**

Reading: *Add Impact with Graphics* 111
Writing: Preparing Graphics 115
Spelling: Double Trouble 125
Vocabulary: Tele/Phono/Photo/Graph/Gram 126
Word Watch: Lose/Lost/Loss/Loose/Loosen 128

CHAPTER 11 **PROCESS REPORTS**   **130**

Reading: *Space Technicians Service the Satellites* 130
Writing: The Process Report 135
Spelling: Getting wISE to IZE/YZE 137
Vocabulary: Micro/Macro 139
Word Watch: Advice/Advise 139

CHAPTER 12 **REVIEW**   **141**

Reading: *Message by Lightwave* 141

PART IV
Business Communications   151

CHAPTER 13 **IN-HOUSE COMMUNICATIONS**   **153**

Reading: *Driving Out the Devils of Communications* 153
Writing: Memos and Forms 156
Spelling/Vocabulary: Seed Roots 167
Word Watch: Past Passed 170

CHAPTER 14 BUSINESS LETTERS 172

Reading: *Turning Confrontation into Communication* 172
Writing: Business Letters 175
Spelling: IBLE/ABLE Endings 186
Vocabulary: GRAD/GRESS Roots 188
Word Watch: Stationary/Stationery Compliment/Complement 189

CHAPTER 15 REVIEW 191

Reading: Ride the Tech Wave or Be Swamped by It 191

PART V
Grammar Units 199

Unit 1   Subjects and Verbs 201
Unit 2   Fragments 210
Unit 3   Compound Sentences 215
Unit 4   Complex Sentences 220
Unit 5   Subject/Verb Agreement 229
Unit 6   Prepositional Phrases 235
Unit 7   Pronouns 240
Unit 8   Pronoun Reference 245
Unit 9   Modifiers 248
Unit 10 Parallelism 256
Unit 11 Avoiding Shifts 261
Unit 12 Avoiding Sexism 267
Unit 13 Transition Words 270

PART VI
Mechanics Units 273

Unit 1 Commas 275
Unit 2 Apostrophes 284
Unit 3 Quotations 288
Unit 4 Other Punctuation Marks 293
Unit 5 Abbreviations and Acronyms 299
Unit 6 Capital Letters 301

APPENDIX 1	COMMON SYMBOLS AND ABBREVIATIONS	305
APPENDIX 2	TIPS FOR TYPING AND WORD PROCESSING	308
APPENDIX 3	TECHNICAL REPORT FORMAT	312
APPENDIX 4	SPELLING AND MISUSED WORDS	323
APPENDIX 5	IRREGULAR VERBS	326
	INDEX	329

PREFACE TO THE TEACHER

This book has been written for writing classes in an electronics curriculum, particularly curriculums in which traditional English books have been ineffective. The primary objective is to provide practical applications of traditional writing skills. Technical students generally present two challenges: they have intense interests in technology, and they have little interest in the language arts. English is not considered a precise science. It is this imprecision that I feel is responsible for the fear and uncertainty experienced by many technical writing students. The solution is to present writing by drawing from their technical subjects.

The approach I have found successful is to use familiar technical vocabulary and concepts to demonstrate less familiar writing strategies and principles, and to encourage students to articulate and organize technical information with precision.

The readings are provided as a springboard into the writing topic and exercises. Each reading could be used to initiate discussions, research, or additional writing assignments. The writing topics are those I consider most relevant for electronics students. At the same time, I have chosen to show the dynamics of language by relating technical language to word derivation, spelling, usage, grammar rules, and general mechanics of writing.

Every writing teacher has personal preferences on the sequencing of various aspects of writing, including the studies of grammar and mechanics. Variations in chapter sequence may be preferred. Four review chapters are provided to reinforce specific skills from the previous few chapters.

I would recommend that the grammar and mechanics units be covered by every student. I have not provided a diagnostic test for this reason. Even though many students are already familiar with some of these skills to some degree, the applications in technical writing are sometimes difficult, even for students who have successfully completed a college composition class. I would suggest assigning one of the grammar/mechanics units each week, to be completed as homework or lab work. These units were designed to be self-teaching and need little explanation.

Preface to the Teacher

The appendices, particularly Appendices 2 and 3, are provided as supplemental work, but merit being assigned as homework, as well.

I invite comments and reactions to the book by you and your students. The Teacher's Manual includes an answer key, some explanations, and sample responses to the exercises.

I have used the *MLA Handbook* and *Webster's New World Dictionary, Second College Edition* as references. Other references may differ on rules, and accepted practices may change. Such is the nature of language.

Andrea J. Rutherfoord

PREFACE TO THE STUDENT

Your electronics skills will be an important factor in getting started in your electronics career. However, if the technical skills of two candidates are equal, the decision for hiring (or promoting) is usually based on the ability to communicate. In some jobs, communications skills are so vital that poor writers or speakers will not be considered, no matter what level of expertise they have in electronics. To confirm the importance of communications skills in electronics, read the "messages from industry" that follow this introduction. Leaders of major electronics firms have offered their personal stands on the value of good communication.

This book was written to teach the types of writing skills you will need to know in an electronics career. You will not write poetry or plays or stories. What you write as you begin a technical career will not necessarily have to be creative or catchy. You will be recounting the facts as you see them; therefore, being able to get the message across clearly and accurately is a survival skill in the technical world.

Before you get started, I want to tell you how to use this book. You will be using a systems approach to communication. A system is an arrangement of related, individual elements that, together, form a unity. Language has several individual elements, as demonstrated in each chapter of this book. Each chapter has five sections. The first section is an article to READ, followed by sections on WRITING, SPELLING, VOCABULARY, and WORD WATCH.

The reading articles have been chosen not only to present examples of good writing but also to provide useful technical or professional information. Too often, students think that facts come only from textbooks (high in information) and entertainment comes from fiction or magazines (high in interest). The articles you will read consist of facts presented in an interesting way. You may encounter new words, and if so, please underline the words and keep reading. Sometimes you will figure out new words by their context, the words around them. Look up the words that you have not figured out after you have finished reading. Then reread the sentences the words were in to see how to use them. Finally, answer the comprehension questions at the end of the article. They were written to help you interpret, organize, and respond to what you have read.

The writing sections deal with one primary writing skill at a time. The skills are those most useful to technical people. In the first half of the book, you will write mainly paragraphs, and in the second half you will write reports and letters. The assignments are similar to the duties of technicians on the job. Get the practice you need now for the types of writing expected of you later.

The spelling sections review some spelling patterns that are reliable. They will also give you a "crutch" or helpful aid to remember correct spellings of tricky words. Poor spelling is sloppy and unprofessional. Since writing takes time, thought, and effort, it seems senseless to degrade our own work with misspellings—it's almost like wearing an expensive shirt inside-out.

The vocabulary sections attempt to bring some order to the haphazard collection of foreign roots, prefixes, and suffixes that form technical words. Latin and Greek, particularly, are responsible for most of the difficult words that we encounter in electronics. These sections will let you practice analyzing words to determine their meanings.

Finally, the word watch sections review groups of easily confused and misused words. Sometimes the placement of one letter completely changes the meaning of a word, as in *tough* and *though*. Other times, two related words, such as *affect* and *effect*, will be studied to determine their correct usage.

Following the chapters, two sections will review the fundamental rules of composing clear and correct sentences: grammar and mechanics. The grammar/sentence structure units deal with the components of language: individual words and groups of words and their functions in communication. These units are necessary to build a foundation for understanding our dynamic and versatile language. Just as electronics is guided by a limited set of theories and principles, English is guided by a limited set of rules. Limited does not mean a small number, just a learnable number. Unfortunately, some students have already given up trying to sort out the rules and their applications. So as you are working your way through these units, you will be asked to examine how words affect communication.

The mechanics units deal with the tools of our language: symbols, abbreviations, numbers, and punctuation. The first and longest unit reviews the comma, which is often perceived as something between a tyrant and a chameleon by many writers. The mechanics units will give you practice using the tools of writing effectively.

At the very end of the book are appendices which offer additional information. Look through them to see what is offered.

Learning the techniques of effective writing requires thought and practice. Whatever effort you put in, however, will pay off—in this course and in your career. Consider this book as a ladder to career advancement. And good luck.

Andrea J. Rutherfoord

ACKNOWLEDGMENTS

This book is the result of many years of learning: from instructors, colleagues, and students. I would like to collectively thank all of them for their individual contributions, large and small.

I would especially like to acknowledge Dr. Joseph Gratto for giving me the opportunity to write this book, Fred Kerr for his technical guidance and suggestions, Carol Lowthian Bartlett for sharing her technical writing expertise, and Greg Burnell from Prentice-Hall for his patient explanations, suggestions, and advice. There are many others who have answered questions and supplied examples, materials, and explanations. Among these are Rodney Coutu (Simplex), Jack Griffin (MDG Systems), Charles Leonard (Scientific Atlanta), Jim Rutherfoord (MRM Associates), and Harrison Whitlow (IBM). My faculty assistants, Chris Hill, Robyn McKnight, and Scott Nantais, provided reliable interpretations of subject difficulty as well as examples of clean technical writing.

My reviewers provided criticisms which sharpened my focus and language. I would like to thank Linda Dobbs, Chris Greveson, Cindy Keller, Stan Kajs, Leanne Murray, Pat O'Connor, Maris Roze, and Scott Schroeder for their diligence and candor.

My greatest appreciation goes to my family. My sons, Chris and Vermont, were my inspiration. They were quietly encouraging, always interested in the status of the book if not in its content. And my husband, Tom, gave generously of his professionalism and energy to read and review the book during all its various stages. His patience and support made the project possible. And so I dedicate this book to them.

To Chris, Vermont, and Tom, with love.

MESSAGES FROM INDUSTRY

Tomorrow's problem will not be the communication of data but more importantly, 'information.' Data abounds. Useful information is still unfortunately sparse.

       Matthew A. Kenny, President
       Racal-Milgo, Inc.
       Sunrise, Florida

The best ideas in the world are useless unless they can be communicated to others. By the word 'communication' I mean that there's not only reception, but understanding of the information conveyed. Miscommunication of information not only destroys many great ideas, but causes untold waste in daily business activities.

       Richard W. Oliver
       Assistant Vice-President
       Nothern Telecom Limited
       Nashville, Tennessee

The ability to speak and write clearly is not only important to the communication of technical concepts, it is an essential part of the innovation process itself. Translating an idea into the written word is one of the better ways of validating the soundness of one's thinking.

       Ian M. Ross, President
       AT&T Bell Laboratories
       Holmdel, New Jersey

Your knowledge is only as valuable as your ability to communicate it to someone else.

> Gerald E. Schultz, President
> Bell & Howell Company
> Skokie, Illinois

...Communications skills are the second most important [skill] that any technical person can learn, and the first is learning to learn. Without communication skills, however, doing the most important is much more difficult than it needs to be. Someone who has already attained communication skills is ready to move ahead more quickly, for he must learn them to move ahead at all.

> Court Skinner
> Manager, Advanced Technology
> National Semiconductor
> Santa Clara, California

One is not evaluated on technical skills alone, but also on the image one presents while communicating.

> Keith R. Welker, Personnel Administrator
> Hughes Aircraft Company
> El Segundo, California

PART I

FOUNDATIONS

Chapter 1 *Getting Started*
Chapter 2 *Completeness*
Chapter 3 *Conciseness*
Chapter 4 *Clarity*
Chapter 5 *Review*

CHAPTER ONE

GETTING STARTED

- *Write formal technical definitions.*
- *Write sentences using the active and passive voices.*
- *Classify words by category.*
- *Write classification sentences with colons and without colons.*
- *Outline a paragraph by finding the main idea and supporting details.*
- *Spell plural words correctly.*
- *Use Latin and Greek number roots correctly.*
- *Use a/an/and and to/too/two correctly.*

TECHNICAL VOCABULARY

Technical writing is different from most of the writing you have done in the past. The writing your future supervisor will expect from you will not be graded for creativity, a large vocabulary, or an individual style. In fact, it will not receive a grade at all, but it will be judged. It will be judged by several different kinds of people: technicians and engineers, supervisors and new employees, co-workers and customers. Some of these people will know more about electronics than you do, some will know less, and some will know very little. They all have one need: they need to understand what you are trying to say. Your writing must be complete, clear, and concise (to-the-point).

1. Complete Good technical writing is informative in nature. It is full of facts organized and presented in a logical, clear, and complete way. Effective communication depends on a full exchange of meaning from writer to reader. Technicians will be asked to write memos, reports, and business letters as well as fill out forms and troubleshooting reports. Leaving out essential information is just as unacceptable as providing misinformation.

2. Clear In these days of information overload (large amounts of data gathered rapidly through technology), your supervisor will be depending on you for understandable sentences, logical ordering of information, and concrete ideas without using a lot of impressive-sounding, abstract words.

For an example of abstract words, look at the following table of words. You will notice that if you choose any word from column A, another from column B, and one from column C, you will create a nonsensical phrase.

A	B	C
relevant	functioning	system
parallel	proportional	subvention
variated	phrased	framework
preempted	logistic	allocation
integrated	optimum	interface

For example, imagine that you overhear someone say, "an integrated (A) proportional (B) framework (C)." Together, the three words sound quite impressive, but in reality, they mean nothing. Be careful that your writing is easy for the reader to understand.

3. Concise Technical readers do not have much time. Often, they would rather be doing something other than reading. If they have to puzzle over sentences or words to figure out what the writer means, they may become frustrated and upset.

Technical writing can be efficient. Writers say things one time only—no repetition, no rewording. Technical terms and *jargon* have been defined and clarified by professionals in your field. Most jargon has been established because it describes an idea or concept in one word. Imagine trying to describe waveforms without using jargon terms such as "sawtooth" or "square." The result would be wordy and cumbersome.

The disadvantage of jargon is that it assumes that the reader also understands the technical meaning of the term. Writing for laypersons, those not expert in electronics, takes special attention. It requires explicit definitions of terms in clear, simple language.

If you suspect that laypeople will be reading your report, you have three choices:

1. Use jargon and hope your reader understands.
2. Avoid jargon and produce a wordy, cumbersome report.
3. Define the jargon you use, and use it freely.

Which is the wisest choice?

DEFINITIONS

Definitions of terms are the foundation of technical writing. A precise set of terms is used in electronics, and only with a common understanding of those terms can information be communicated accurately.

Some terms have entirely different meanings in electronics from those with which you are familiar in everyday life. Examples of such words are "power," "force," and "communication." For example, the term "communication" used in casual conversation can include speaking, listening, reading, writing, and body language. But to an electronics technician, if the message wasn't transferred electronically, it wasn't communicated at all. In fact, the study of communications systems begins with Samuel Morse's invention of the telegraph in 1837, even though we all know that throughout history, people have been sending verbal and nonverbal messages to anyone who would pay attention.

Some terms are used with more precision in electronics than in everyday life. Words such as the following have precise meanings in electronics and must be used carefully:

absolute	ground	rate
critical	intensity	relative
current	inversely	specific
force	potential	static
fundamental	power	uniform

Some terms are frequently confused. Can you state the difference between "force" and "power"? These are words that are used interchangeably every day, but in electronics the meanings are different. There are many such terms in technical writing.

Sometimes students are in such a hurry to do their homework problems that they skip to the end of the chapter, referring to the chapter only as a last resort. Don't be this kind of reader. Eventually, you will have to communicate what you know in words, either spoken or written. You won't be able to communicate entirely in numbers. Get used to how the experts, the authors, describe the principles of your technology. Learn the terms and how to use them correctly. Several examples of short and long, formal and informal definitions are presented in this chapter to give you practice in this skill.

INFORMAL DEFINITIONS

You can probably remember learning your first definition in electronics.

Resistance: opposition to current flow.

This is an informal definition. A definition placed between commas or parentheses is usually an informal definition.

A *potentiometer* (variable resistor) is used for volume controls.

If too many informal definitions are used, a report may become disjointed and distracting. Normally, a writer who plans on using more than two unfamiliar technical terms in a report will define the terms formally in the introduction or glossary.

FORMAL DEFINITIONS

A formal definition has two functions: it identifies the larger class that the term belongs to, and it provides distinguishing characteristics:

term > class > characteristics

The *term* is the word being defined, for example, "Porsche." A Porsche is in the *class* called automobiles, or more specifically, German automobiles. The definition goes on to provide *distinguishing characteristics* or details about the term that make it different from other members of the group.

A Porsche, pronounced "pour-sha" (*term*), is a German-made automobile (*class*) with high-performance capabilities, a small, aerodynamic body design, and a price tag starting at $25,000 (*characteristics*).

A formal definition can be written for any electronics term, and often the most difficult part is determining the class! For example, is resistance a quality, a capability, or an action? Technicians must occasionally make such subtle distinctions, and it is worth some thought.

Device	*Property*	*Capability*	*Process*
an object	a trait	an ability	an action
resistor	resistivity	resistance	resist

In which of the groups above do the following terms belong?

resonance	resonant
resonator	resonate

Resonator is an object, something you can touch, so it would be in the class of devices. *Resonate* is an action verb, so it is a process. *Resonant* is how we describe an object, an adjective, so it is a property or trait. *Resonance* is the capability of performing the action.

Once the group has been determined, technicians usually don't have much trouble furnishing the distinguishing characteristics.

A resistor   is an electronic device
(term)   (class)

that is used in electronic circuits to oppose and control current flow. Its capacity to resist current is indicated by a color code or stamped values.
(distinguishing characteristics)

One final point to remember is to avoid using the term, or any variation of the term, in the remainder of the definition.

Wrong: A *resistor* is an electronic device that *resists* current flow.

In the example above, find a *synonym* (another word with the same meaning) such as "oppose" or "control."

DICTIONARIES

Dictionaries give varying types of definitions. Smaller, pocket-sized dictionaries rarely offer technical definitions. Look up "resistance" in a small dictionary, and you will probably find only the root word, "resist," with several common endings, but no mention of current flow.

A "technical dictionary" will offer only electronics-related definitions. One example is *The Illustrated Electronics Dictionary* by Howard Berlin (Charles E. Merrill Publishing Company, 1986). These useful guides are popular with beginning electronics students and people unfamiliar with technical terms.

Complete definitions for general and scientific uses are found in a "collegiate dictionary." If you look up "resistance," you will find up to seven or more distinct meanings, defining how the word is used in different situations. Keys such as "Elec." or "Physics" indicate the specific definition used in electronics technology.

Writing technical definitions is the first skill that you will practice as a beginning technical writer.

Reading Comprehension Questions

1. What are the three characteristics of good technical writing?

2. What is jargon? Why is it important?

3. What are the three parts of a definition?

4. What is the difference between a formal and an informal definition?

5. What information can you find in a collegiate dictionary?

6. Look up "frequency" in two dictionaries, one small and one large. Copy the entire entry from each dictionary, and underline the meaning that you use in electronics.

WRITING: Getting Started
- Definitions
- Active and passive voice
- Classification
- Writing lists
- Noting organization

DEFINITIONS
You have just learned that the primary characterstics of a formal definition are *term, class,* and *distinguishing characteristics.*

Note When writing a definition, after stating the term, do not use it again in the remainder of the definition.

> *Wrong:* A power supply supplies power to a circuit.
> *Correct:* A power supply is an energy source used to operate electrical and electronic devices.

Exercise 1.1 Write formal definitions for the following terms. For the first five, classes are provided. If necessary, use a dictionary. Write the definitions as though you were beginning a discussion with nontechnical people (a group of high school students, for example).

1. Power supply (*device*) _____

2. Proton (*particle*) _____

3. Lead (*wire*) _____

4. Closed circuit (*current path*) _____

5. Decible (*unit of measure*) _____

6. Capacitance _____

7. Conductor _____

8. Transformer _____

9. Power _____

10. Dielectric _____

Write a formal definition of your school. Make your definition complete.

ACTIVE AND PASSIVE VOICE

In speaking, you can vary your voice and tone to communicate a message more effectively. For example, speaking slowly and clearly with a strong voice will add authority to a message. Speaking in a lively voice with changes in pitch and tone can add interest to the message. Good speakers know how to use their voices to inform, persuade, or entertain an audience.

In writing, we also have different voices. The most common are called *active* and *passive*. In **active** writing, the subject acts on an object.

> The technician *placed* the resistor on the breadboard.

Active verbs make the sentence personal and lively.

In **passive** writing, the object is stated first. In the following example, notice how the emphasis has been changed from the subject to the object.

> A resistor *was placed* on the breadboard by the technician.

Notice, too, that passive writing requires more words and does not have the zip of active writing.

> *As a rule of thumb, write in the active voice as much as possible. But in formal technical writing, it is sometimes effective to write in the passive (impersonal) voice. Remember that poorly worded passive writing can sound artificial, awkward, and wordy.*

Although recent trends in technical writing show increasing use of the active, personal style, it is good to be familiar with the passive style. Use it when you want to emphasize the object rather than the subject.

Exercise 1.2 Rewrite the following sentences in the passive voice. To help you, an outline of the answer has been given with each question, and the verbs you will need to change have been put in italics.

Example

ACTIVE VOICE: We *know* this series as a Fourier series.
(This series . . . as. . . .)
PASSIVE VOICE: This series *is known* as a Fourier series.

1. Mr. Noel *checked* your circuit design.
(Your . . . Mr. Noel.)

2. They *considered* many technical and economic factors.
    (Many . . . were. . . .)

3. Heat from the transformer *burned* some of the insulation.
    (Some . . . by heat. . . .)

4. We *obtained* an increase in reliability by using transistorized circuits.
    (An increase . . . can . . . by. . . .)

5. We *found* a quick way to solve the equation.
    (A quick way to . . . was. . . .)

Now change the passive sentences to the active voice.

6. The fundamental definitions are stated in words and formulas.

7. A cycle of alternating charge and discharge current was provided by a capacitor.

8. The amount of capacitance was calculated by the technician.

9. The effects of the rate of change in a sine wave are illustrated in the example.

10. More details are explained in the next chapters.

CLASSIFICATION

Well-written technical articles present information in an order that eliminates the need for repetition. Once a term is defined, it can be used freely. Each section of the article discusses only one idea. Key ideas are underlined or titled. Ideas are presented in an order that prepares the reader for the next idea. It is this ordering that keeps technical writing "to the point."

Technical writers plan the logical ordering, or organization, of information before they start writing. They often use a strategy called **outlining,** which is a numbered and lettered sketch of the contents. They may then use the outline headings and subheadings as labels within the article or report.

Grouping or classifying information is a method that some people use for memorizing information. Try to memorize the following list of words.

amplifier	fuse	NOR
AND	giga	NOT
capacitor	kilo	OR
conductor	mega	tera

Now try to memorize the same words in a different order.

Metric Units	*Devices*	*Gates*
giga	amplifier	AND
kilo	capacitor	OR
mega	conductor	NOR
tera	fuse	NOT

The first list is in alphabetical order, and the second list is in categorical order. It is often easier to remember items grouped by categories. When we organize, or group information in categories, it is also easier to discuss or write about the information.

10 *Getting Started*

Exercise 1.3 Rearrange the following words into categories (four words per category) and label each category.

	Ampere	chips	joules	henries
	tubes	wired	scopes	tested
	Coulomb	ohms	watts	soldered
	solved	multimeters	Ohm	Kirchhoff
Labels:	_____	_____	_____	_____
	1.	1.	1.	1.
	2.	2.	2.	2.
	3.	3.	3.	3.
	4.	4.	4.	4.

A paragraph, or a whole report, will usually have one main idea, such as a category, and several supporting details, such as the items within a category. Sometimes an author will state the category and list the details within the category, as in the following sentence.

> There are three methods of analysis that use Kirchhoff's laws: branch current analysis, mesh current analysis, and node voltage analysis.

WRITING LISTS

A list at the beginning of a paragraph, section, or report is an effective way of telling the reader how you have organized your ideas. The second sentence in the example above states the list in an obvious way by using a **colon** (:) before the items. Although either way of stating the list is correct, the colon is a visual cue and thus is more direct.

The colon is a punctuation mark that means "as follows." It is a signal to the reader that a list is being given. Other signals for lists are words and phrases such as the following:

There are three kinds:	. . . can be	divided into three areas:
types:		classified
groups:		grouped
classes:		classed
categories:		categorized

Example: Resistors can be divided into two groups: fixed and variable.

There are two types of current: alternating and direct.

A colon is usually not used following a verb. After a verb, simply write your list.

Example: The two main categories of resistors are fixed and variable.

If your list includes more than two items, put a comma between the items. If commas are needed within items, separate the items with a **semicolon** (;).

Example: The article can be divided into two sections: the specialized and multipurpose information utilities; and the requirements of the hardware, computer-communications network, and software.

For typing rules about these punctuation marks, see Appendix 3.

Exercise 1.4 For each of the following outlines, write a sentence that names the category, signals a list, and lists the items in the category. You are asked to write three sentences using a colon, and two sentences without using a colon.

Example

    A. Resistors
       1. Fixed
       2. Variable
       Sentence: There are two categories of resistors: fixed and variable.

1. Engineering
   a. Electrical
   b. Mechanical
   c. Civil
   Sentence (use a colon) _____

2. Transistor circuits
   a. Common base
   b. Common collector
   c. Common emitter
   Sentence (use a colon) _____

3. Uses of integrated circuits
   a. AF and RF amplifiers
   b. Gates, flip-flops, and additional types of logic circuits
   Sentence (use a colon) _____

4. Basic gates
   a. AND
   b. NAND
   c. OR
   d. NOR
   Sentence (do not use a colon) _____

5. Types of radio receivers
   a. AM (amplitude-modulated)
   b. FM (frequency-modulated)
   Sentence (do not use a colon) _____

NOTING ORGANIZATION

When you take notes on a reading or a lecture, you can use an outline to simplify complicated subjects. Notes not only force you to organize spoken or writ-

Getting Started

ten information (which is necessary for memorizing), but they become a valuable tool when reviewing for tests.

Because an outline is a shortened version of what you have heard or read, you will not have to write as many words. Writing an outline is a way of selecting the main ideas and important supporting details. Practice and careful reading or listening will improve this skill.

Exercise 1.5 In each of the following paragraphs, a category will be given, followed by at least two details. Write the outline of each paragraph. Remember to be brief, accurate, and complete.

Example: Switches are commonly used to open or close a circuit. Closed is the ON, or make, position; open is the OFF, or break, position. The switch is in series with the voltage source and its load. In the ON position, the closed switch has very little resistance. Then maximum current can flow in the load, with practically no voltage drop across the switch. Open, the switch has infinite resistance, and no current flows in the circuit.

  Main idea: Uses of switches
  Details: a. Closed, ON, or make position—low resistance
       b. Open, OFF, or break position—high resistance

1. A battery is a combination of cells. After discharge, a primary-cell battery cannot be recharged because the internal chemical reaction cannot be restored. After it has delivered its rated capacity, the primary cell must be discarded. A secondary-cell battery can be recharged because the chemical action is reversible. Because it can be recharged, it is also called a storage cell.

   Main idea: _____

   Details: _____

   _____

2. Most electronic circuits can be classified into three groups: rectifiers, amplifiers, and oscillators. A rectifier changes its ac input to dc output. An amplifier circuit amplifies its input current. An oscillator circuit is a special case of an amplifier, but the oscillator generates an ac output from its dc power supply, without any ac input signal.

   Main idea: _____

   Details: _____

   _____

3. The bipolar transistor refers to either an *NPN* or a *PNP* transistor. The first section supplies charges, either holes or electrons, to be collected by the third section, through the middle section. The electrode that supplies charges is the emitter; the electrode that collects charges at the opposite end is the collector. The base in the middle forms two junctions between emitter and collector, to control the collector current. This type of transistor is also called a BJT (bipolar junction transistor).

   Main idea: _____

   Details: _____

   _____

4. Failures in transistors generally result from an open weld at the wire leads to the semiconductor, a short circuit at a junction caused by momentary overloads, or circuit failures that cause transistor overheating. In most cases, a defective transistor is internally short-circuited or open, and simple tests will reveal the trouble.

Main idea: _____
Details: _____
_____

5. The name "flip-flop" refers to the ability of the circuit to remain in one of two stable states. The action is like that of a latching relay. An input pulse sets one state for the flip-flop. It will remain that way until the next input pulse resets the flip-flop to the opposite state, which is also stable. The two opposite states are considered stable conditions because an input pulse is needed to change the level in the output.

Main idea: _____
Details: _____
_____

SPELLING: Plurals

PLURAL NOUNS

Nouns can be **singular** (one) or **plural** (more than one). Normally, to change a word from singular to plural, we simply add a final S to the end of the word. If the singular word ends with S, X, Z, CH, or SH, we add ES to make the word plural.

one noun	two nouns
one switch	three switches
a tube	many tubes

There are several exceptions, however, that show up frequently in writing. In this section, we review some of the **irregular plurals** that you may encounter. The one sure way of spelling a plural word correctly is to look the word up in a dictionary. Usually, an irregular plural ending will be noted early in the entry with *pl.* followed by the last syllable of the plural spelling. If two spellings are given, either is correct and the first is preferred. For example, the entry for "deer" will look similar to the following:

deer (dir) *n., pl.* deer, occas. deers

This tells you that "deer," pronounced as "dir," is a noun, and the plural spelling is "deer" or, occasionally, "deers."

If you look up a noun and no plural is given, you can assume that the plural is regular and just add a final S.

Y PLURALS

If the word ends with a Y, change the Y to I and add ES.

| frequency | two frequencies |
| country | many countries |

However, if the final Y is preceded by a vowel (a, e, i, o, u), as in *delay*, do not change the Y. Changing the Y to an I would result in three vowels next to one another. We don't do this in English, because it makes words difficult to read; we simply add an S to the end.

Wrong: delaies
Correct: delays

14 *Getting Started*

Another exception to this rule relates to the plural form of proper nouns. Do not change the spelling of the noun, and do not add an apostrophe. Just add a final S.

>two Jerrys   all the Rileys

In the case of abbreviations, numbers, letters, and **acronyms** (words made from one or more letters each of several words in a term or phrase). To make a plural form, simply add an S or add an apostrophe-S after the last letter.

>several IBM PCs   two 1's
>many M.D.'s   all CRTs

———————— **NOTE** ————————

We sometimes add an S to the end of verbs, also. If the verb ends in Y, the same spelling rules as those noted above apply when changing the verb to its "s" form.

>We carry   He carries
>We say   He says

Exercise 1.6 Make each word plural.

1. array _____   2. memory _____
3. display _____   4. carry _____
5. specify _____   6. priority _____
7. Henry _____   8. henry _____
9. theory _____   10. vary _____

F PLURALS

Usually, when a word ends with F or FE (wife), we change the F to a V and add ES.

>a wife   all wives

The common exceptions to this rule are as follows:

>one belief   many beliefs ("believes" is a verb)
>one brief   many briefs
>a handkerchief   many handkerchiefs
>a roof   many roofs
>a safe   many safes ("saves" is a verb)
>one sheriff   many sheriffs

Use the dictionary to be sure.

Exercise 1.7 Make each word plural.

1. The _____ of writers differ.
       ability

2. He reshingled _____ in his spare time.
           roof

3. The speaker introduced himself to the _____.
                         employee

4. The men escorted their _____ to the banquet.
   wife
5. The spreadsheet allowed several _____ of data.
   array
6. The entire department consisted of two _____.
   sheriff
7. The two former _____ of staff spoke at the luncheon.
   chief
8. There are three _____ in our department.
   Johnson
9. Several _____ have been done on time management.
   study
10. There are many storage _____ on the market today.
    device

HYPHENATED WORDS
For hyphenated words the S or ES is added to the main word in the compound.

sister-in-law   sisters-in-law
son-in-law   sons-in-law
vice-president   vice-presidents
cross-examiner   cross-examiners

IRREGULAR PLURALS
Some words have structural changes in the plural form. This is especially true of foreign words and endings.

child   children
foot   feet
woman   women
medium (in communication)   media
datum/data (either is allowed)   data
alumna (female)   alumnae
alumnus (male, mixed)   alumni
syllabus   syllabuses/syllabi

SIS TO SES
Adding another ES would be a tongue-twister.

crisis   crises
thesis   theses
diagnosis   diagnoses

Exercise 1.8 Write the correct plural.

1. The two _____ brought their _____.
   daughter-in-law   *child*
2. The writer could not decide between his two _____.
   thesis
3. After graduation, a reception was held for all the _____.
   alumnus
4. We were alerted to all the _____ through the _____.
   crisis   *medium*

Getting Started

5. The _____ were collected automatically.
   datum

6. She was introduced to each of the _____ of operations.
   vice-president

7. Troubleshooting reduced his original _____ to one.
   diagnosis

8. His back ached after standing on his _____ all day.
   foot

9. Mr. Striker handed out _____ in all his classes.
   syllabus

10. Both _____ questioned him thoroughly.
    cross-examiner

VOCABULARY: Latin and Greek Number Roots

LATIN NUMBER ROOTS

Latin number roots are found in many technical words. Most people can recognize these root words by themselves, but it takes observation to find them in other words.

Latin Root	Number	Example	Meaning Using Latin Root
Uni	1	unique	one of a kind
du	2	duplex	two-way communications
tri	3	tricycle	three-wheeled cycle
quara	4	quarter	one-fourth
quint	5	quintet	five-piece musical group
sex	6	sextet	six-member group
sept	7	September	seventh month of old Roman calendar
oct	8	octopus	eight-armed sea creature
nov	9	novena	nine-day religious devotion
dec	10	decimal	number expressed in base ten
cent	100	centigrade	thermometric scale from 0 to 100 degrees
mill	1000	millimeter	1/1000 of a meter

Exercise 1.9 Use the Latin number roots to complete the correct words.

1. A number system to the base eight is called an _____ al system.
2. A _____ al-slope A/D converter uses a variable- and a fixed-slope ramp.
3. An exam given every three weeks is called a _____ -weekly.
4. The number system to the base ten is called the _____ imal system.
5. One-thousandth of an ampere is called a _____ iampere.

Exercise 1.10 Write an informal definition of the following Latin-based words. Use complete sentences.

1. Century *Example:* A century is a 100-year period.
2. Dual-purpose _____
3. Decathlon _____
4. Trilateral _____
5. Octogenarian _____

6. Quadrant _____
7. Unilinear _____
8. Sexennial _____
9. Million _____
10. Quintuplicate _____

GREEK NUMBER ROOTS

Many Greek number roots are used in everyday English. Since ancient Latin and Greek are sister languages, some numbers will be identical or similar to the Latin number roots.

Greek Number	English Number	Example	Meaning Using Greek Root
mono	1	monosyllable	one syllable
di	2	dioxide	two parts oxygen
tri	3	tricycle	three-wheeled toy
tetra	4	tetrachloride	four parts chlorine
penta	5	pentagon	five-sided figure
hexa	6	hexameter	six beats per line
hepta	7	heptarchy	government of seven rulers
oct	8	octopus	eight-armed sea creature
ennea	9	(this root is almost unused in English words)	
dec	10	December	tenth month of the old Roman calendar
hecto	100	hectogram	100 grams (metric)
kilo	1000	kilometer	1000 meters (metric)

Exercise 1.11 Use the Greek number roots to complete the sentences.

1. The number system to base 16 (six plus ten) is called _____ imal. (Use two roots)
2. Laser light is _____ chromatic since it produces only one color.
3. A _____ watt is a 1000-watt unit of energy.
4. A situation with a choice between two unpleasant alternatives is a _____ lemma.
5. The first five books of the Bible are called the _____ teuch.

Exercise 1.12 Write an informal definition using the Greek root. Write complete sentences.

1. Monostable _____
2. Dialogue _____
3. Trilogy _____
4. Octant _____
5. Hexagon _____
6. Heptavalent _____
7. Pentode _____
8. Tetragon _____
9. Kilovolt-ampere _____
10. Decade _____

WORD WATCH: A/An/And
To/Too/Two

USING A, AN, AND

The words A, AN, and AND are often misused. Sometimes this happens because the writer is not sure when to use A or AN. Other times, the writer means to say AND, but forgets to add the final D. These three words are *not interchangeable*. They each have a specific function, or job, in sentences. Using the wrong word will make you appear careless.

A/AN are used in the same situations (in front of single nouns), but one *or* the other is used.

a car	an orange car
a circuit	an integrated circuit
a direct current	an alternating current
a slope	an elevation
a uniform	an hour
(uniform begins with	(hour begins with
a hard U or Y sound)	an O sound)

Do you notice a pattern? Look at the first *sound* following A/AN. Notice that sometimes the first *sound* of the word is different from the first *letter*.

Use A before words beginning with a consonant sound.
Use AN before words beginning with a vowel sound.

Remember that **vowels** are the letters A, E, I, O, U (and sometimes Y). All other letters are **consonants**. Notice that the first *sound* of a word is not always the first *letter*.

AND is a conjunction used between words. Sometimes people do not pronounce the final "D" or fail to add it when they write. Remember that AN is not a substitute for AND.

──────── **Rule** ────────

Use A in front of a word beginning with a consonant sound.
Use AN in front of a word beginning with a vowel sound.
Use AND to join words.

Exercise 1.13 Fill each blank with A, AN, or AND.

Time management is _____ skill that students must learn if they hope to handle all their responsibilities. Reducing traveling time saves _____ hour here _____ there. Bringing _____ bag lunch means being able to eat without hunting for _____ inexpensive cafe. These things are necessary for people trying to go to school, keep _____ part-time job, study, _____ most important, find some time to relax. Many students find _____ hobby that is inexpensive _____ active. Jogging, calisthenics, _____ dancing are _____ few activities that provide _____ healthy outlet for the stress that builds from keeping _____ demanding schedule. With good pacing and planning, students can make the most efficient use of their time _____ keep _____ enthusiastic, energetic attitude.

USING TO, TOO, AND TWO
TWO is the number (2) and is rarely confused with TOO or TO.

> There were only *two* hours left.
> There could be *two* interviews.

TOO is a modifier and can have TWO meanings. It can mean "more than enough" when it is in front of another modifier:

> The gray jacket was *too* expensive.
> The blue tie was *too* long.

TOO can also mean "also." When used this way, TOO will usually have commas around it:

> I'll interview with that company, *too.*
> They, *too,* have positions available.

TO is used in all other cases, usually as a direction (preposition) or in front of a verb.

> I selected the suit *to* wear *to* the interview.
> I planned *to* arrive early.

Exercise 1.14 Choose the right form of TO/TOO/TWO for each sentence.

1. I arrived only _____ minutes early for the interview.
2. I gave my résumé _____ the receptionist and introduced myself.
3. She said it wouldn't be _____ long before she would take me _____ the manager.
4. I noticed that she made _____ phone calls.
5. Suddenly, a door opened and a woman asked me _____ step into her office.
6. She introduced herself _____ me as the personnel manager.
7. I was _____ nervous _____ remember her name.
8. Fortunately, her nameplate was sitting just _____ feet in front of me.
9. Unfortunately, it gave just her first and last names, and I didn't know whether _____ call her "Miss" or "Mrs.".
10. I used the title "Ms." and I was able _____ say her name _____ times during the interview.

CHAPTER TWO

COMPLETENESS

- Outline paragraphs by finding the main idea and supporting details.
- Write a paragraph with a main idea and supporting details.
- Spell words ending in ly correctly.
- Use negative prefixes correctly.
- Use wear/were/we're/where correctly.

READING: Wardrobe: The First Step to Your Professional Image

SUSAN BIXLER

Computer Dressing

In the early seventies the "science" of business dress was born. The emergence of "computer" dressing brought the unofficial uniform look, which was very similar for both men and women: a navy suit with a long-sleeved white shirt. It was not bad as a first concept but it overlooked individual jobs, personalities, and especially personal taste. "Formula" dressing was just plain boring.

The corporate uniform may have been a necessary first step toward an accepted professional look, but it was a monotonous one. It became obvious that no single look could successfully be pasted on everyone, like one set of cut-out clothes on millions of paper dolls. Clearly, a professional image could not overlook individuality.

Fortunately, the eighties have seen the softening of the business uniform, especially for women, most of whom simply will not be bullied into wearing a dark, tailored suit every single day. Women are secure enough to wear dresses with and without jackets. Men, too, have learned that a navy-blue three-piece suit might be an excellent wardrobe choice, but it certainly isn't the only one.

The same power uniform actually loses its punch when it is worn every day. It becomes so commonplace that it goes unnoticed. Certainly, consistency in dress is important, but the power uniform—the heavy artillery—is usually best saved for occasions or situations that demand it.

Besides, such a "power" uniform, whether it is a masculine or

"Wardrobe: The First Step to Your Professional Image," from *The Professional Image* by Susan Bixler. Reprinted by permission of The Putnam Publishing Group. Copyright © 1984 by Susan Bixler.

feminine version, can work against you if it is worn inappropriately. It may intimidate people. After all, professional dress should not serve to make you outshine everyone else, nor to "one-up" a client. *Image is a tool one uses to accomplish an objective.* A casually dressed retail owner in a rural area is not going to be comfortable trying to do business with a power-suited "city slicker."

The clone look is dead—from overexposure. It gave no one an advantage and therefore no one has any reason to mourn its passing.

Dressing for Your Audience

Rita Jenrette may have helped lose the case for her Congressman husband, John, because of her courtroom attire. While he was pleading bankruptcy, she was wearing thousand-dollar designer outfits. His verbal testimony lost credibility because of her visual message. Contrast this to the almost matronly look that high-fashion model Cristina Ferrare adopted when her husband, John De Lorean, faced a long series of court appearances immediately after his narcotics arrest. She was aware of how her image could affect his verdict. So she opted for dark, very simple suits, a short plain hair style and a small amount of makeup and jewelry. She adapted her appearance to her new circumstances and to a different audience.

It was not entirely the weather that prompted former New York City Mayor John Lindsay to shed his suit jacket and tie several years ago when he was trying to win votes in the city's low-income neighborhoods. He was attempting to make himself more appealing to his audience. This principle is exactly what professional dressing is all about.

If your audience is the board of directors, you dress for them. If it's the guys at the loading dock, you take off your jacket, roll up your sleeves, and shed your tie or scarf. Varying your wardrobe will vary your impact and serve you better. And remember that every outfit delivers a message.

A woman who shows up for a job interview in a lemon-colored Ultrasuede suit looks as though she should be lunching at the country club, not launching a career. A man who shows up for the same interview in a shiny polyester suit, a short wide tie, and a sweater vest does not appear to be taking the interview seriously.

The woman is saying that she doesn't really need to work, that she is a dilettante looking for something to do between tennis matches. The man's casual look is saying that the job is unimportant.

Good business people are always conscious of the immediate impression they make with their attire. A man trying to sell a small computer to the guy who owns the corner Shell station is probably going to be making his sales presentation while the prospective buyer is in greasy overalls. An aware salesperson would feel foolish standing there in a three-piece suit clutching an attaché case. The gas station owner would feel intimidated. A sensitivity to the selling situation means that the salesperson will shed his coat, roll up his sleeves and loosen his tie before strolling into the station. However, if his next call is on the vice-president of sales at a Fortune 500 company, the tie is put back into place, and the suit jacket is brushed off and put on.

Smart Choices

Appropriate is the key word. And within the appropriate range there are usually sufficient choices available to satisfy yourself, your company, and your customers. When I consult with individuals, I am more successful when I help them determine what works best for them, rather than simply dictating stringent lists of do's and don'ts. For example, a veteran engineer for a computer company was promoted to head of his division. He wanted some help with his wardrobe and needed an opinion about his nicely trimmed beard. Although it was thick and well groomed, I had some reservations. I asked to see the people in his department and found that most of the other men had beards, too. But I advised my client to shave off his so that he wouldn't look like "one of the guys." He needed a separate, more corporate identity. The engineer looked handsome in a beard, but his career opportunities and future advancement were more enhanced when he was clean-shaven.

Similarly, a female branch manager for a small bank wanted to know whether it was all right to wear a blond wig on days that her hair looked dirty. Instead of simply telling her to wash her hair more often, I asked her the same question that I put to the engineer: "Will it help or hinder your career advancement?" She made the decision to adopt a simpler hair style that would be easier to maintain and would eliminate the need for the wig.

Each of us has the opportunity to develop whatever type of visual message we wish to send to our bosses, co-workers and clients. It is possible to dress above or below our actual level in business. Many ambitious young people dress above their positions, consciously trying to create the impression that they are destined for better things. If the image is backed up by ability, they usually are.

If you work for a large corporation, pay attention to the way the successful employees dress. You will probably notice that people's dress improves as their importance within the company increases—it rarely goes in the other direction. Top-level management people tend to look successful. You would not often mistake a junior clerk for a senior vice-president; if you did, that might well indicate the clerk was on the way up or the vice-president on the way down.

On the other hand, our office received a call when a fast-tracked woman was promoted to a higher level and gradually began turning in her conservative skirted suits for a fashionable bangled bracelet look. Her supervisor called our offices to request an individual wardrobe consultation. After a discussion, it was apparent that the woman knew how to dress professionally. She was simply electing not to do so. This newly promoted woman was indicating visually that her present position was as high as she wanted to go. She was intentionally, or perhaps unintentionally, telling the company not to consider her for future promotions.

Manipulating Your Appearance

Occasionally a situation demands dressing down. A successful Southeastern developer finds it works to his advantage, in some instances, to dress the part of the "good old boy," especially when he is outside the corporate limits of Atlanta and Birmingham. His clients are then farmers or builders wearing hard hats. We suggested that he not wear blue jeans and boots—but no banker's pinstripe either. In this situation, a navy blazer, open-collared, button-down shirt and a pair of pressed khaki pants work nicely.

A female journalist says it sometimes works to her advantage to be understated, depending on the person she is interviewing. In her "power" costume—dark suit and tailored blouse—she notices that certain people tend to be more cautious about what they say; dressed more casually—a dirndl skirt and a well-tailored business sweater—she puts people more at ease and they often tell her more than they initially intended. But it is her audience that determines her attire. When she spoke at a monthly meeting of female journalists, she wore a much stronger outfit.

By manipulating our clothing and appearance, we are using visual impact to its best advantage. We can ascribe an importance or lack of importance to ourselves, and give a strong visual suggestion of our background, education, and future prospects.

Getting Paid for Looking the Part

Research indicates that it is financially beneficial to present a dynamic, well-polished image in business. Funded by the Clairol Corporation, psychologist Dr. Judith Waters from Fairleigh Dickinson University researched the impact an effective business appearance has on a starting salary. She sent out a large number of "before" and "after" photographs and identical resumes to more than a thousand companies. No company received both a "before" and "after" picture. Each was asked to determine a starting salary.

The results were amazing. They indicated an initial salary of eight to twenty percent higher as the result of upgrading a mediocre business appearance to one that is crisp and effective.

Employers are quite willing to pay more for people who already look the part. There is the inference that employees who care about themselves will care about their jobs.

If employees are already projecting an image of professionalism through their dress, that's one less thing—and a potentially unpleasant thing—that the firm has to worry about. It also leaves more time for instructing the employee on product information and company procedure, and for other essential training.

American businesses and corporations will generally reward, through increased salary and promotions, those professionals who project effectively. A well-polished image gives you a genuine competitive edge and stamps you as someone on the way up.

Reading Comprehension Questions

1. What does "power uniform" mean?

2. How will you know what clothes are appropriate working clothes at a company?

3. What type of situation would require "dressing down?"

4. What things do you think affect first impressions? Describe what they tell you about a person.

5. What things do you do differently when you are trying to impress someone?

WRITING: Topic Sentences

In Chapter 1 you learned how to use definitions and details to describe objects. Both definitions and descriptions are important parts of paragraphs. They add the precision and clarity that are necessary to relay technical information. But the most important element of a paragraph is the topic sentence.

The *topic sentence* states the main idea of a paragraph. It sets the limits of the paragraph. It can appear at the beginning, middle, or end of a paragraph—or not at all (the inferred topic sentence). Good technical writers usually start each paragraph with a clear topic sentence to ensure readability and logic.

Consider the topic sentence, *Resistors are divided into two groups: fixed and variable.* You can easily predict that the content of the paragraph will be limited to details about the two types. Perhaps the paragraph will describe one type, the first one listed, and a second paragraph, beginning with "The second type . . . ," will complete the idea. Identifying the topic sentence is an important task for the reader.

For practice in paragraph analysis, outline the following paragraphs. The first main idea has been given to you.

Exercise 2.1 Write the supporting details from the paragraph.

COLOR POWER. Men are more restricted than women in their color choices. Some colors, green, for example, detract from a business look. Solid brown is dangerous; it might work on an older man when used in a rich-looking fabric in just the right warm hue, but usually it has either a lower-class or a country-squire look. Black is the ultimate power color; yet for men, it is overpowering in anything but formal wear. Even though they seem to come "in" periodically, shirts in lavender, light green, and tangerine are not good business choices. Generally, it is best for men to stick to more conservative colors.

Main idea: Men are restricted in their color choices.
Specific details:

1. _____
2. _____
3. _____
4. _____
5. _____

Notice how each detail contributes to the main idea.

24 Completeness

Exercise 2.2 For further practice, outline the following paragraphs, listing the main ideas and the details.

COLOR DECISIONS. Whether you opt for stripes, plaids, or prints, the guiding precept for any pattern is subtlety. Men and women may use small, subtle plaids, such as glen plaids, for suits, but a plaid suit jacket must never be paired with a pair of trousers of another design, because the combination will not look right. Herringbone is a very subtle, classical pattern that is both elegant and sophisticated. In heavier fabrics, it is a good choice for blazers for both men and women. In lighter fabrics, it is an enduring suit pattern. Stripes can work well to offset a dark suit. Very thin pinstripes can enhance the authority of a navy or gray suit. Stripes should always be narrow and vertical. Prints are best used as small dots and foulards—the classical geometric Ivy League look. A classical print should never be larger than the head of a pencil. Women may wear prints in a blouse or dress, and men can choose them in a tie.

Main idea: _____

Specific details:

THE CLASSICAL LOOK. The classical cut, styling, and detail last forever. The classical look doesn't change every season, and it always looks correct. A man's suit with three-inch lapels, natural shoulders, and straight-legged trousers has a classical look. Men's classic suits are two-button, three-button, or double-breasted. A woman's suit in a conservative color with slightly padded shoulders and a permanent hemline will not go out of style. Women's classical suits have fully lined jackets with long, set-in sleeves. Classical skirts may be straight, dirndl (slightly gathered but not full), A-line, or pleated. High-quality classical clothes may cost a little more, but they are definitely cost-effective because of their durability and versatility.

Main idea: _____

Specific details:

Now it is time for you to write your own paragraph. You should begin with a topic sentence or main idea. Use the rest of the paragraph to explain your topic through details and description.

The exercise below provides a form for writing a response to Susan Bixler, the fashion consultant. Use one of the following topic sentences for your letter, and go on to describe the "uniform" or "power suit" in detail. The purpose of this assignment is to state a topic sentence and provide only enough details to support your statement.

> The "uniform" of students is quite different from that worn in a business.

or

> I have already noticed how a "power suit" has helped my self-confidence in important situations.

Exercise 2.3 Write a one-paragraph response to the author of the article beginning with one of the main ideas listed above. Provide specific details to support your main idea.

_____
_____
_____
_____
_____
_____
_____
_____
_____

Reread your paragraph and correct the spelling, wording, grammar, and punctuation. Also notice whether you supported your main idea. If so, you have written a complete paragraph.

SPELLING: Adding LY and LLY

One way to turn an adjective into an adverb is to add LY to the end of the word. Sometimes writers are confused about whether to add LY or ALLY. A dictionary will always provide the answer, but there are some general rules that you can follow.

_____ **RULE** _____

Add LY to the end of an adjective to make an adverb.

Add LY to the end of the following adjectives to turn them into adverbs.

fortunate + ly = _____
serious + ly = _____
physical + ly = _____

Most of the time, if a word ends with a final, silent E, just add LY. There are four common exceptions.

Exception 1: In these words, drop the final E before adding LY.

simple + ly = simply
due   + ly = duly
true   + ly = truly
whole + ly = wholly

Exception 2: If the word ends in BLE, drop the final E before adding LY.

terrible + ly = terribly
sensible + ly = sensibly

Completeness

Exercise 2.4 Change the following adjectives into adverbs by adding LY.

1. horrible _____
2. awful _____
3. especial _____
4. precise _____
5. accurate _____
6. true _____
7. probable _____
8. whole _____
9. seasonal _____
10. simple _____

Exception 3: If the word ends in a final Y, change the Y to I before adding LY.

easy + ly = easily
heavy + ly = heavily

But if the final Y sounds like I, just add the LY.

dry + ly = dryly
sly + ly = slyly

Exception 4: If the word ends in C, add ALLY.

medic + ally = medically
critic + ally = critically

(In the case of PUBLIC, add LY = publicly.)

Exercise 2.5 Complete the spelling of the adverb form.

1. The mayor public_____ announced her intention to run for reelection.
2. The child played happy_____ in the yard.
3. Solving the problem was simple_____ a matter of rewording the question.
4. The new hairstyle changed his appearance drastic_____.
5. We were terrible_____ shocked when Tim walked out of the interview.
6. The rain started, fortunate_____, before the crops died.
7. The passengers were critic_____ injured in the accident.
8. Steve finished the test easy_____ within the hour.
9. Graduation is true_____ an exciting experience.
10. With your positive attitude, you are like_____ to succeed.

VOCABULARY: Negative Prefixes

There are several negative prefixes that we use to change a word to the opposite or to signify "not."

Negative Prefix		Root Word		New Word
a	+	symmetric	=	asymmetric
de	+	activate	=	deactivate
dis	+	appear	=	disappear
il	+	legal	=	illegal
ir	+	relevant	=	irrelevant
mis	+	placed	=	misplaced
non	+	sense	=	nonsense
un	+	known	=	unknown

Note Do not change the spelling of the root word when adding a prefix.

Exercise 2.6 Choose the correct negative prefix.

Use MIS or DIS.
1. Theo used his calculator so that he would not _____calculate.
2. We stared at the spectacle in _____belief.
3. The _____orderly appearance of the lab was corrected.
4. The technician corrected the _____spelled words.
5. Susanne knew that leaving work early would _____please her manager.

Use IR or IL. (Notice that the beginning letter of the root word sometimes gives a clue to the prefix.)
6. Some of the damaged equipment was nearly _____replaceable.
7. The results of the experiment seemed _____logical and _____rational.
8. The assignment was _____legible because the pencil lead smeared.
9. The _____regular towels were unusual shapes and sizes.
10. The _____literate man could not read the documentation.

Note The prefix ILL (usually hyphenated, ILL-) means bad.

> *Examples:* an *ill-fated* experiment
> an *ill-natured* boss
> an *ill-advised* procedure

Use UN or NON. (Often UN- and NON- are used interchangeably, but one will be used more commonly.)
11. This section of the restaurant is reserved for _____smokers.
12. My supervisor was _____willing to let us slip our deadline.
13. The happy-go-lucky worker seemed _____troubled and relaxed.
14. The American Cancer Society is a _____profit organization.
15. Michael finally realized that his _____professional behavior cost him a promotion.

Use A or DE.
16. A right triangle and an equilateral triangle are _____symmetrical.
17. The radio receiver had to _____code the message.
18. As the car approached the stop sign, the cautious driver started to _____celerate.
19. Since the clocks were _____synchronous, the machines could not communicate with each other.
20. The radio would not work because a transistor was _____fective.

Note These prefixes *do not always* mean "the opposite of," so be careful when using them.

> *Examples:* He read the passage *aloud* (out loud).
> The American dollar has been *devalued* (reduced in value).

WORD WATCH: Wear/Were/We're/Where

Use WEAR as a present-tense verb referring either to clothing or overuse:

> I'll be careful about what I *wear* to the interview.
> My old blue suit shows a lot of *wear* and tear.

Use WERE as a past-tense verb:

> We *were* told to consider gray or blue.

Use WE'RE as a contraction for "we are."

> *We're* willing to spend several hours with a tailor.
> or
> *We are* willing to spend several hours with a tailor.

Use WHERE in referring to a place:

> I don't know *where* to find a suit for under $100.

Exercise 2.7 Use WERE or WHERE.

1. We decided to shop in a store _____ tailoring was available.
2. We _____ willing to pay a little more to get the right fit.
3. The clerk pointed to the racks _____ we would find three-piece suits.
4. He then told us _____ to find the fitting room.

Use WE'RE or WEAR.

5. The tailor advised us to _____ black or gray stockings.
6. _____ still surprised at the attention he gave each of us.
7. Now that we have our business suits, _____ more confident about interviewing.

Use WE'RE or WERE.

8. During the next few months, _____ planning on dressing like professionals.
9. We _____ used to wearing only jeans and T-shirts.
10. Now _____ certain we project an image by how we dress.

Exercise 2.8 Use each form correctly in a complete sentence.

1. (WHERE) _____

2. (WERE) _____

3. (WE'RE) _____

4. (WEAR) _____

Have your answers and sentences checked by your instructor.

CHAPTER THREE

CONCISENESS

- *Write a summary.*
- *Spell words with* ie/ei *correctly.*
- *Use* spec/son *roots correctly.*
- *Use* they're/their/there *correctly.*

READING: Shedding Light on Today's Lasers

ROBERT LEE HOTZ

Laser scanners tote up the groceries at checkout counters, help the Central Intelligence Agency read the world's newspapers, and trace the outline of Georgia in the night at Stone Mountain.

The laser, 25 years old this spring, can generate beams of light precise and powerful enough to drill holes in diamonds, weld the retina of a human eye into position, or perform microsurgery on individual cells.

Tuned to the right frequency and unleashed at high power, beams of laser light can knock out heat-seeking missiles moving at 2,000 miles an hour. In more benign form, they animate the games in a video arcade.

Since its invention, the laser (which stands for light amplification by stimulated emission of radiation) has beamed its way into medicine, defense, communications and computers, even art and entertainment.

"I can't think of any concept that has had a bigger impact in 25 years," said Dr. James L. Gole, a professor at Georgia Tech's School of Physics, where lasers are used to explore high-temperature physics. "I can't think of any device that is even close. It has opened up many areas that before were impossible to probe."

In May 1960, Dr. Theodore H. Maiman built the world's first laser at Hughes Research Laboratories. With a rod of synthetic ruby, he harnessed light to radiate a high-energy beam of a pure red light.

Maiman's laser excited electrons into emitting photons—the basic unit of light energy—at a single wavelength, then pumped the pure, "coherent" light into an amplified beam as red as the ruby that produced it.

"The laser ranks up there with the vacuum tube, the transistor and the integrated circuit," says Dr. Bernard S. Finn, curator of "The

"Shedding Light on Today's Lasers," by Robert Lee Hotz, from *The Atlanta Constitution,* June 6, 1985. Reprinted by permission of The Atlanta Constitution.

SHEDDING LIGHT ON TODAY'S LASERS

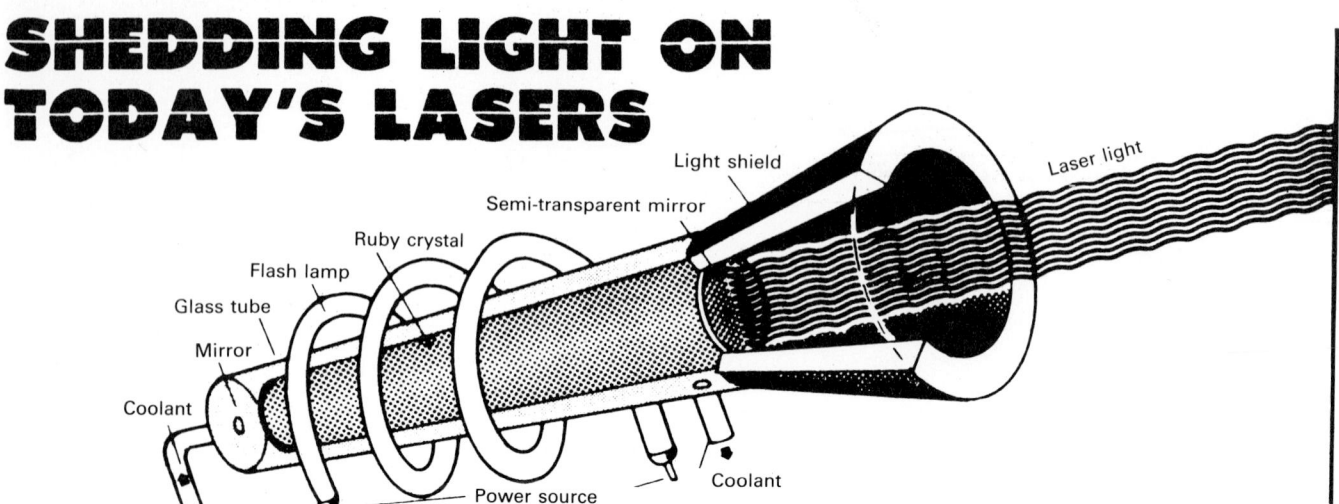

Sunlight is a mixture of colors of many different wavelengths, each moving in its own way. The waves are out of step, and this is why the edges of the colors in the spectrum of sunlight are fuzzy (as in rainbows). Red light waves move in step, but they are not parallel. The spectrum of red light is also fuzzy at the edges. Red laser light waves are in step and parallel. The spectrum of red laser light is clear and sharp at the edges.

Laser at 25," an exhibit organized by the Smithsonian's National Museum of American History to observe the invention's anniversary. Over the next three years, the exhibit is scheduled to travel to 15 cities, including Huntsville, Ala., in November.

Maiman's first laser generated only red light, but today lasers come in as many different colors as they do shapes and sizes.

The world's most powerful laser—10 amplifiers, each longer than a football field—fills a large building at the Lawrence Livermore National Laboratory in Livermore, Calif. The smallest—a semiconductor 350 microns wide and 250 microns long—is small enough to float on a teardrop.

The Livermore laser, named Nova, took eight years and $176 million to build. Its predecessor—a carbon dioxide laser called Antares—generated more than 12 trillion watts of radiant energy in a single burst lasting one-billionth of a second. Dedicated last month, Nova will be used in weapons research and to study thermonuclear fusion power.

Tiny semiconductor lasers are the heart of grocery store cash register scanners, sophisticated communications systems, optical radar and a new breed of computer that one day may help orbiting defense systems detect and track incoming missiles.

At Georgia Tech, a team led by Dr. Tom Gaylord is working with scientists from eight universities as part of a $9 million Star Wars research project aimed at developing a computer that can perform 1,000 simultaneous calculations. Conventional computers can handle only one operation at a time.

Gaylord's group is developing a computer memory that uses a laser to store information as a three-dimensional image in a synthetic crystal. The technique vastly increases available computer memory and provides instant access to it. A conventional computer must sort through its stored data one bit at a time.

To test one element of potential Star Wars technology, scientists in Hawaii later this month will fire a laser beam at a reflector aboard the space shuttle Discovery. Pentagon reseachers hope to learn how much the light spreads on its way through the atmosphere and how to correct it, knowledge critical to the construction of a large laser weapon.

Coming down to Earth, laser technology is no farther away than most wallets and pocketbooks. Visa and MasterCard, plagued by counterfeiters, incorporate glittering three-dimensional laser images called holograms into their new credit cards to foil forgers. Each is etched on a thermoplastic square only millionths of an inch thick.

Lasers have worked their way into many facets of society:

- As "light scalpels," they have transformed the science of eye surgery. They are also used to cauterize bleeding ulcers, remove tumors, tattoos and birthmarks. Doctors now use lasers to clear clogged arteries and to do delicate neurosurgery.

- As surveying instruments, lasers have been used to keep tunnels true, calculate the speed at which the continents drift apart, and measure, to the nearest foot, the distance from Earth to the moon. Lasers can now accurately measure the spinning nucleus of an atom.
- As weapons, lasers still generate little more than promise, but government scientists last month reported an important advance in the development of a laser space weapon that could be powered by a nuclear explosion. The secret X-ray laser at Livermore, called Super Excalibur after King Arthur's magic sword, reportedly uses a new technique to focus the explosion and generate the powerful energy beams. The device was tested this spring in an underground explosion in the Nevada desert.
- As transmitters, semiconductor lasers can send billions of bits of information a second through hair-thin strands of glass. Fiber optic systems already carry telephone conversations and television signals over limited distances, and a proposed trans-Atlantic "lightwave" cable—4,719 miles of glass fiber now being manufactured by Western Electric in Norcross, will carry 40,000 simultaneous telephone conversations when operational at the end of the decade.
- As recorders and storage devices, individual laser discs can store and play back up to 100,000 images. They are used in automated teaching systems for industry and the armed services. In the home, laser-operated video discs compete with video tape recorders and laser-powered compact disc players compete with traditional turntables and tape decks. Experimental computer memories use laser discs to store information at densities 100 times the capacity of conventional magnetic discs.
- As optical sensors, lasers process the images for robots and missile guidance systems. Scientists at Carnegie-Mellon University built a system for the CIA that scans thousands of microfilmed articles for pages of special interest. It replaced dozens of desk-bound human analysts.

"Many of the applications of lasers that were foreseen haven't come to fruition," says Michael O. Rodgers, senior research scientist at Tech's School of Geophysical Sciences, who builds laser sensors to detect trace elements in the atmosphere.

"But I don't know of anyone in the laser business 10 years ago anticipated a time when large numbers of lasers would be used in people's homes for compact disc players and video discs.

"The degree to which fiber optics and laser communications have progressed has been incredibly fast," he said. "That technology has matured enormously, to the point where it is quite competitive with communications satellites."

1983 Laser & Laser System Sales by Application

Application	Laser System Sales ($ million)	Laser Component Sales ($ million)	Example
Nonimpact printing	805	29	High-speed printers
Optical communications	530	14	Fiber-optic communications
Color separation	505	9	Four-color printing
Tactical military	309	54	Range finding
Metrology, alignment	265	17	Leveling
Video disc, audio disc	222	7	Video, audio recording
Medical	196	41	Diagnosis, therapy
Materials processing	143	65	Cutting welding
R&D	97	84	Laboratory instrumentation
Point-of-sale scanning	76	2	Supermarket check-out
Typesetting, platemaking	34	2	Newspapers
Entertainment, display	4	3	Light shows
Total	3176	317	

Reading Comprehension Questions

1. What is the difference between laser light and sunlight?

2. Explain the meaning of the following quote:

   "The laser ranks up there with the vacuum tube, the transistor, and the integrated circuit."

3. Define a photon.

4. What examples are given of the largest and smallest lasers at the time the article was written?

5. Describe three applications of lasers with which you are familiar.

WRITING: Writing a Summary

A **summary**, sometimes called an **abstract**, is a condensed account of the essential information included in a longer piece of writing. An abstract is placed at the beginning of an article or report. A summary appears at the end. The function of a summary or abstract resembles that of a schematic diagram, which gives a clear, brief presentation of a circuit without the clutter of the actual materials necessary to build the circuit.

For example, if you needed information on a report about lasers, you would find information in many sources, too many sources to actually read. You may find professional abstracts of journal (magazine) articles. By reading these brief summaries, you would be able to judge which articles would be most useful to you.

A summary answers the basic questions that readers want answered before they devote more time to reading the article or book. Many people who are interested in keeping up with technology do not have the time to read every article printed about their field. They often rely on professional abstracts to find the most useful articles and disregard the rest.

A reader searching for information has predictable questions for each article:

WHAT? WHO? WHERE? WHEN? WHY? HOW?

The answers will be found in professional abstracts, although they may not be in that order. Only the key ideas and conclusions will be included.

Follow these helpful steps when writing a summary.

1. Read the article carefully—more than once—before starting to write. Use your pencil to mark key ideas, phrases, and conclusions.
2. Look for the author's own summaries at the beginning or end of the article. Often, boldface headings indicate a transition and a new key idea.
3. Note the author's organization—find the main idea of each paragraph or section.
4. The length of a summary is usually about 33 percent of the length of the article, although this is by no means a rule. Instructors seldom require more than one page, and professional abstracts are rarely longer than one paragraph, no matter how long the article.
5. Summarize each section (of longer articles) or paragraph (of shorter ones). Disregard figures of speech, examples, detailed descriptions, and discussions.

6. Do *not* include personal interpretations, agreements, or disagreements (no "I" statements). Write in the third person (he, she, it, they).
7. Read the article once more and compare it to your summary. Make any revisions that are necessary for clarity.
8. *Format:* the summary should contain the following information.
   a. Identification of the article being summarized (name of author, title of article, title of book or magazine, date of publication).
   b. Statement of the main idea of the article.
   c. Statements that explain all the important points used to support the main idea.
   d. Explanation or clarification of important points, if necessary.

The laser article is summarized below. Note how the ideas from the article are reworded and condensed. Copying exact sentences is considered plagiarism. Do not quote other people's writing in either an abstract or a summary.

> "Shedding Light on Lasers," written by Robert Hotz (*The Atlanta Constitution*, 6/11/85) provides a 25-year history of the laser and its applications. The first laser (light amplification by stimulated emission of radiation) was built in 1960 at Hughes Research Laboratories by Dr. Theodore H. Maiman, who discovered how to form a single-wavelength, high-energy beam of pure red light. Lasers are now built in many shapes and sizes and produce a variety of colors. The laser is currently being used in tests of the proposed "Star Wars" project and by credit-card companies to prevent forgeries. Lasers are also frequently used as "light scalpels," surveying instruments, weapons, transmitters, recorders and storage devices, and optical scanners.

Notice that the purpose of the article, the main idea, is stated first. Then the key word "laser" is defined. Notice, also, that the last section of the article can be summarized in one sentence since each paragraph (labeled with a "bullet") simply describes an application.

Exercise 3.1 Write a one-paragraph summary of the reading article from Chapter 1 or 2. Limit your summary to no more than 10 sentences.

TITLE:
_____
_____
_____
_____
_____
_____
_____
_____
_____
_____

Show your summary to your instructor.

SPELLING: IE/EI

You probably remember learning the verse that handles most of the IE/EI problems.

*I before E,
Except after C.*

believe	conceive
achieve	receiver
relief	deceit

*Or when sounded like A,
As in neighbor and weigh.*

sleigh	eight
their	freight
neighborhood	weight

As you would expect, there are some exceptions to these rules. The most common exceptions are listed below, and you will have to remember them. Notice that the exceptions are all EI spellings.

either	neither
seize	seizure
weird	leisure
counterfeit	forfeit
foreign	height

Do not confuse the EI/IE vowel pairs with "unrelated" combinations, such as in *science* or *reinforce*, in which both vowels have a distinct sound, thus a logical order.

Exercise 3.2 Add EI or IE to spell each word correctly.

1. Credit-card companies have been plagued by counterf_____ters.
2. It was a rel_____f when the Nevada test results y_____lded positive results.
3. N_____ther sc_____ntists nor engineers predicted the applications of lasers in l_____sure and recreation industries.
4. Dr. Maiman has rec_____ved acknowledgment for his ach_____vement.
5. Lasers have worked th_____r way into many facets of soc_____ty.

Exercise 3.3 Proofread the following paragraph for IE/EI errors. There are six errors. Write the correct spellings above the incorrect words.

Passing through an airport customs' line in a foreign country can be a wierd experience. Travelers can expect that officials may sieze suspicious-looking items for examination. If the traveler has purchased anything illegal, even unintentionally, he or she will have to forfeit the item. Freinds patiently prepare each other for thier inspections. It is always a releif to see each peice of luggage pass through an inspection.

Exercise 3.4 Write the EI/IE verse from memory. Make sure that you include the second part.

VOCABULARY: Roots SPEC and SON

In the description of lasers, you read about the *spectrum* of sunlight and laser light. The root of *spectrum* is the Latin word *spec* (*spect*), which means "look." Literally, *spectrum* means "visible light waves."

The Latin root *son* means "sound." Combined with the suffix IC, meaning "having to do with," we form the word *sonic,* meaning "having to do with sound."

Exercise 3.5 Using the root words SPEC and SON, complete the words.

1. A person who watches or observes is called a _____tator.
2. An instrument to aid vision is called a pair of _____acles.
3. A musical composition written in three or four movements is called a _____ata.
4. A remarkable sight that attracts onlookers is called a _____acle.
5. Several people singing one melody are said to be singing in uni_____.
6. A close examination of an item is called an in_____tion.
7. A descriptive statement issued by a new company is called a pro_____tus.
8. A noise out of harmony is described as being dis_____ant.
9. A device used to increase vibrations is called a re_____ator.
10. A thought or conjecture formed from thinking about various aspects of a subject is called a _____ulation.

Exercise 3.6 Using the Latin meanings of SPEC and SON, write definitions of the following words *as they apply to technology*. Use the dictionary if necessary. Remember to define the words by using the term, class, and characteristics.

1. specifications (specs) _____

2. spectrometer _____

3. supersonic _____

4. sonic boom _____

5. resonance _____

WORD WATCH: They're/Their/There

THEY'RE is the contraction of THEY ARE. If a substitution of "they are" is logical and appropriate, use the contraction form.

> *They're* the people I told you about.
> (*They are* the people I told you about.)

THEIR is a possessive. It is followed by a noun.

> It is *their* turn.
> We drove to *their* house.

36 Conciseness

THERE is an adverb often used at the beginning of sentences (but it is not a subject) or a noun that indicates location, direction, or time, the opposite of here. You can see the shorter word HERE, which provides a clue for when to use the word.

There are two answers. (*adverb use*)
Put it down over *there*. (*noun use*)

("Here" could be substituted in both sentences.)

Note All three forms are spelled beginning with T-H-E.

Exercise 3.7 Complete the following sentences with the correct form of THEY'RE/THERE/THEIR.

The first time I attended a meeting of the Robotics Club, I watched the other members closely. I was interested in _____ projects and goals. I sat _____, not intending to speak, when I was called on to introduce myself. After standing _____ speechless for a few seconds, I finally mumbled my name and quickly sat down. Now, after attending several meetings, I realize that _____ more knowledgeable about robotics than I am, but that _____ also interested in my ideas. It's not just _____ club, but it's my club, too.

Exercise 3.8 Write two sentences using each spelling.

THEY'RE (use as the subject and verb of the sentence)
1. _____
2. _____

THEIR (follow with a noun)
3. _____
4. _____

THERE (use as a direction or the beginning of a sentence)
5. _____
6. _____

Check your sentences with your instructor.

CHAPTER FOUR

CLARITY

- *Correct verb errors.*
- *Correct redundancy errors.*
- *Write a paragraph comparing something new to something familiar.*
- *Complete word analogies.*
- *Double the final consonant correctly when adding a suffix.*
- *Use* sub/super *prefixes correctly.*
- *Use* used/supposed *correctly.*

READING: Taking the Noise Out of Technical Writing

JOHN W. McDONALD

In any communication system, the accuracy and intelligibility of the message is limited by noise. Noise enters written communication as "linguistic noise" (misfit words, confused syntax, and faulty organization) and "typographic noise" (printing errors, poor graphic design, and faulty reproduction). The concepts of noise filters, modulation, and signal-to-noise ratio are applied to technical writing in terms of careful writing, editorial monitoring, and relevance of the message to a particular audience.

"Taking the Noise Out of Technical Writing," by John W. McDonald from *Technical Editing: Principles and Practices*, May 1975, p. 28–32. Reprinted by permission of The Society of Technical Communication.

Introduction

Noise has recently become a topic of great interest. Unwanted and excessive noise pervades our society, with effects ranging from the distractive to the destructive. Legislative bodies have responded to public awareness of the detrimental effects of noise by imposing standards for industry, construction, and transport. One wonders whether the sonic boom of the SST will ever be as loud and disturbing as the political noise it has already generated.

All communication, with the possible exception of divine revelation, takes place in the presence of noise. In all person-to-person communication, we must rely on our truly remarkable ability to detect useful signals in an environment of ubiquitous noise. And hope that we don't make matters worse by irritable reaction to the task of sorting out what is useful in a badly constructed message that is garbled in transmission, distorted by the carrier or medium, and filtered through our own sieve of preconceptions and prejudices.

Before we talk about how to reduce noise in communication, let's discuss what noise is, and how it gets into the communication system.

Sources of Noise

The basic limitation on the speed and accuracy of any mode of communication—whether oral or written—is due to noise that distorts the message, with consequent confusion for the receiver. For this talk, we will take the usual physics textbook definition: noise consists of unwanted disturbances, especially those that are random and persistent, that obscure or reduce the clarity and quality of a signal.

A simplified communication model consists of (1) the source, (2) the message, (3) the channel, and (4) the receiver, all superimposed on a background of ambient noise. A fifth element, that of feedback, establishes a link between the source and the receiver so that we can instantly evaluate how well the receiver is getting and understanding the message and so that the sender can modify or reinforce his message by reducing noise or by modulating the channel. Here we take modulation to mean any process that varies the signal for more efficient transmission, with the intent of superimposing intelligence on, or increasing the information content of, a carrier wave. The concept of modulation as applied to the editorial process may give rise to a quite different sort of noise if an author thinks his "style" is being violated.

In face-to-face conversation, noise comes from mispronounced words, distracting mannerisms, esoteric vocabulary, inattention, and such environmental effects as traffic, air-conditioner hum, and the like. To partly overcome interference, man's language is highly redundant so that many words and sounds in speech can be lost without serious effect on the message being sent. Redundancy allows man to detect signals in noise, and intelligibility can be achieved even when the noise intensity is very much larger than the message sound. This is described as the "cocktail party effect," as many of you will remember from the Wednesday night reception, whereby conversations can be carried on in an extremely noisy room. But just barely.

In conversation, we can use the instant feedback of a question, a raised eyebrow, a puzzled grimace, a rebuttal, or a glazed eye to alert the sender that the message isn't coming through.

All person-to-person communication suffers from noise, some more than others, but none so much as writing. The elements of the written communication system are (1) the writer, (2) the report, and (3) the reader. The system lacks an instant feedback channel that links the reader with the writer, and the first transmission has to be clear—and clearly understood. The reader has no way to influence the quality of the message except by angry "letters to the editor" long after the fact, and the writer cannot be certain that he is not creating noise.

In written communication, noise can occur at each of the three elements.

The writer can contribute linguistic noise at his semantic encoder. He may become the victim of his own semantic inadequacy that allows misfit words, mangled metaphors, and deadwood to creep in. The report itself may contribute typographic noise by printing errors, faulty reproduction, cluttered illustrations, and poor graphic design. Finally, the reader may be the victim of psychological noise, the source of which may be the message itself, semantic or typographic noise in the report, or some external or subjective effect.

Noise Abatement at the Source

Technical reports, by their nature, resist being measured by a common standard. But if there is any common measure for judging technical writing, it is organization—and faulty organization is most readily revealed by comparison with an outline.

Outlines. Writers fall into two classes: those who find outlines useful or indispensable, and those who consider them a nuisance. If you're in the first category, use an outline as a starting point and in-process guide; if you're in the second class, use an outline anyway but prepare it after the report is completed as a *post hoc* verifier.

Outlines have no standard form; they must be variable and flexible to suit the subject and scope of coverage. Nevertheless, an outline is as essential to solid writing as blueprints are to construction. The final product should not reveal the skeleton; the writing should be strong enough to hold the report together and solid enough so that no bare bones poke through. Minimum requirements for an outline are:

- State the epitome of the subject and your attitude about it.
- Give evidence to support your thesis.
- Summarize what you want the reader to remember, with sufficient support from evidence and sufficient clarity of expression to keep him from asking "So what?".

Another noise filter applied at the source is to remember that it is impossible to "tell all." The author must make a selection of data and arguments, but the selection must fairly correspond to the mass of evidence and it must offer a graspable design to the beholder.

Carelessness. The commonest source of noise is not the willful distortion of intent, but innocent lapses of attention that jerk us to wakefulness. It requires the unnodding vigilance of writers and editors alike to keep us from taking sweeping statements with "a dose of salts" rather than "a grain of salt."

Our daily work abounds with good examples of bad habits, all sources of noise. We have to know what correct usage is, which depends in part on knowing the difference between what a word denotes and what it connotes. We must draw fine shades of difference between *accuracy* and *precision*, between *activate* and *actuate*, between *alternate* and *alternative*. The list is long.

We must also watch for mangled metaphors that result from trying so hard to reach the descriptive high notes that the singer dies of syntactical strangulation. We all have lists of boners gleaned from our editorial labors. Some of my favorites are:

- "A virgin field pregnant with possibilities."
- "The need is evident for a list of physicists broken down by specialization."
- "No authenticated case is known in which sterile parents transmitted this quality to their offspring."
- "This publication fills a much-needed gap."

Our attention is demanded in culling out tautological repetition of the same sense in different words, as in "general consensus of opinion." And we must prune deadwood—it's as burdensome to prose as it is to a tree.

Someone may object that the distinctions are too finely drawn. Not so. Strictly speaking, technical communication is a matter of speaking strictly.

Spelling. A word is more than the sound it makes. It is, among other things, the way it looks on a page—and it looks best when it's spelled right. An author who writes with no misspelled words has already dismissed the first grounds of suspicion of the limits of his scholarship. He might remember that the fancy way of saying "correct spelling" is "orthography" (from the Greek roots *orthos* meaning *straight,* *correct, normal;* and *graphein* meaning *to draw, to inscribe*).

A good dictionary is a communicator's most useful single tool. An author who frequently consults a dictionary will usually learn something to his—and his reader's—advantage.

Proofreading and Goof-proofing. Examine the finished product to be sure that the message cannot be misinterpreted. Substantiate your claims, check your data, and ask colleagues for their opinions. If you are uneasy about your data or insecure in your interpretation, your writing will show it. And proofread! Proofreading is equivalent to goof-proofing. Look up words whose meanings have been dulled by familiarity, get the typos out of the final draft, and watch out for grammatical lapses. Don't be hasty to rush your literary brainchild to a baptism of printer's ink.

Editorial Modulation

Only rarely in creative activity, whether in research, in writing, or whatever, can we achieve a desired result at the first try. Rarer still is a draft manuscript that goes from writer to editor to printer without retracing part of its path for correction or improvement.

The act of creating any written piece starts a process that resembles the oscillation of a vibrating spring; this concept enables us to follow the path of a manuscript on successive trips between author and editor (see Fig. 1). Imagine that the initial displacement from the solution axis is proportional to the degree of uncertainty, and that the restoring force of a spring represents the process of evaluation and generalization. At the maximum amplitude of each oscillation we apply a "damping" factor that consists of the act of defining and refining, and thus limit the next oscillation. Successive definitions at peak amplitudes, as the draft goes from author to editor and back again, contribute to decreases in the degree of ignorance. The envelope formed by decreasing displacements of these quantities is the convergence of understanding. It approaches, at infinity, the truth.

At the Los Alamos Scientific Laboratory, we construe editorial modulation to mean editing for consistency, organization, and language correction, with the specific

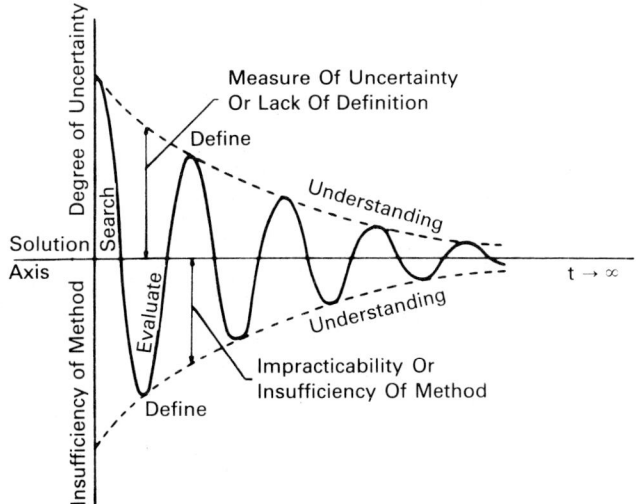

FIGURE 1 The creative process as a damped oscillation.

intent of retaining the author's style and the accuracy of technical content. Some writers complain that an editor's efforts have changed the intended meaning. When that occurs, the writer was probably not clear in the first place. A writer who is concerned lest editorial modulation violate his "style" can best preserve it by

- Writing clearly and directly.
- Making sure of his facts and marshaling them in logical sequence—with the "news" of his findings or a statement of the problem first, followed by details and supporting arguments.
- Imparting the significance of his work to readers outside his field.
- Providing enough details of method and equipment so that others can repeat his work.

He should apply a noise filter at the source by avoiding overspecialized jargon and acronyms, by using the active voice in preference to the passive, and by using personal pronouns (I, we) where appropriate. "The present author" went out with "gentle reader," so say "I" naturally and honestly when it fits. "We," unless the work is reported by more than one author, is used only by royalty, editors, and people with a tapeworm. Finally, it is no sin to occasionally split an infinitive or to use a preposition to end a sentence with.

Noise Abatement in Report

Noise filters applied to the reproduction of a report are easier to use and control than the human factors of source and receiver. We can, by consistent and conscious effort, filter out printing errors, faulty reproduction, cluttered illustrations, overcrowded text, inadequate margins, stiff bindings, misplaced references, and poor display of mathematical expressions. The filter of language correction will take care of most lapses in punctuation, orthography, transitions, unity, consistency, and conformance to "house" style.

It is another matter, however, to nurture a happy editor-author symbiosis wherein writers accept the suggestions of editors who understand the limitations and requirements of printshop and audience, wherein editors learn the subject matter from authors in a mutual search for clearer writing, and neither insists on having his own way.

Noise Abatement at the Receiver

When we're at the receiving end of the communication chain, psychological noise may arise from misinterpreting the sender's intent, or from listening only to what we want to believe. Aside from this irreducible residuum of human cussedness, we know that the linguistic and typographic noise of a report, added to noise from the environment or from the recesses of our subconscious, can become psychological if it's bad enough. Our impatience with an author who seems not to be meeting his obligation of clear exposition can contribute to psychological noise. Noise filters at the receiver must, by and large, be applied by the receiver himself if he's willing to understand and takes the effort to seek and tune to the author's frequency.

Optimizing Signal-to-Noise Ratio

I guess it's possible for one person's noise to be another person's signal, but it is the relationship of what is relevant to what is extraneous that must be our constant concern. We, as communicators, are obliged to be source, message, channel, receiver, and feedback at one time or another—and sometimes we are asked to be all of them at once. To my mind, then, we must

- Have something to say (optimize the signal).
- Say it concisely and simply; avoid adjectival overkill (apply noise filters at the source).
- Say it in terms of interest to the reader (minimize noise at the receiver).
- Write correctly so that editors cannot distort the message or suppress the style (optimize editorial modulation; reduce noise in the report).
- Adjust the emphasis or choice of words as necessary to make sure that the message is getting through (pay attention to the feedback signal).
- Try to inform, not impress. Write as we would be written to.

Writing for Understanding

It has become a commonplace to say that scientists can't write. This is not true. Many of them write creditable prose every day, and some of them write very well indeed. To say that scientists can't write is an insult to those who do, and discourages those who should. Nevertheless, the quality of many manuscripts submitted to our Department for publication indicates that research scientists regard the recording and reporting of their work as an unpleasant addition to their already heavy burdens. Nothing of the sort. Writing, rather than impeding research, may properly be regarded as the mechanism by which the work is accomplished.

Michael Faraday, the great British physicist and lucid expositor of his own work, is reported to have defined research as consisting of three steps: Start, finish, publish. Research, in this view, starts on the basis of well-made plans that define a specific field of interest and finishes when results are obtained that are of a degree of accuracy compatible with the experimental design necessary to prove the hypothesis. Results are then written and distributed in a clear exposition of observations made and results obtained. I would encourage the writing of laboratory notes in daily increments as a stimulus to the scientist to think about his work in a more organized way and to use the intellectual discipline of writing as daily feedback to the actual progress and shape of the work.

Some complicated concepts are difficult to explain in an uncomplicated way, but the intellectual challenge requires an attempt to solve puzzles, not create them. The challenge lies not only in achieving scientific and technical advances, but also in meeting the demand for conveying involved concepts clearly and concisely; additional effort is required to present good ideas effectively. Scientists from different disciplines, in their attempts to explain something to a third party, may become comprehensible to each other. It is not enough to be understood; we must make sure that we are not misunderstood.

Reading Comprehension Questions

1. What is "noise" in writing?

2. a. How does the writer contribute noise?

   b. How does the report contribute noise?

   c. How does the reader contribute noise?

3. What are some methods for eliminating noise at the source?

4. What are some methods for eliminating noise in the report?

5. What are some methods for eliminating noise at the receiver?

6. Why do you think it is important for technicians to write for understanding?

WRITING: Slang, Clichés, Analogies

Read *slowly*.

Mah favorite drinks are Coke, Sebun-up, and Dagapeppa.

Yew're ownie young wunst, but if yew know whut yore doin', wunst is enuff.

Officah, thet thar ve-hicle was goin' slap down the middle a' the dang road.

Waycheer! Ah'll just wandah over yawnduh and cum to attair cha-yuh.

In Chapter 2 you read that the clothing people wear affects how others judge them. The way we speak is also judged. Listeners will judge not only WHAT you say, but HOW you say it. If HOW you say it is unconvincing (careless, awkward, unprofessional), the listeners may not pay attention or try to understand.

Sometimes slang or regional dialects are entertaining and colorful. Some dialects are so different that we wonder if the people speaking them are speaking English at all! Unusual expressions can sound friendly and quaint.

Occasionally, slang is used intentionally to understate a person's education ("Ah'm jist an ole country lawyah"). Many young people have a street language that they use with their friends ("Hey man, wha's happ'nin"). Dialects are appropriate with certain groups of people, at certain times.

The type of English used in textbooks and the professional world is called **standard English**—free of slang and dialects. Speaking and writing standard English is sometimes like using a second language. People who want to project an educated, successful image realize that speaking nonstandard English is like wearing a flashy orange suit—everyone notices and few forgive.

Some people may blame their nonstandard English on the educational system or other background differences. The truth is that many people have overcome the grammar barrier simply by concentrating not only on what they say but on how they say it. Let's face it, if you want professional people to take you seriously, you must talk (and dress and act) like a professional.

VERB ERRORS

The most common speaking offenses usually involve verbs. A speech habit of leaving off the final D, EN, or S will in turn cause writing problems, where the error becomes even more noticeable. Pay attention to the final letters on words.

Another common error is using *be* as a verb by itself. In standard English, *be* is only a helping verb and must have another verb with it. Also, when we speak fast, we sometimes fail to pronounce helping verbs, especially if we are using a contraction. Be sure to include them in writing.

Wrong: Don't use that machine. *It be broke.*
Wrong: Don't use that machine. *It's broke.*
Correct: Don't use that machine. *It's broken.*

Wrong: He *don't* know the answer. *I seen it on his face.*
Correct: He *doesn't* know the answer. *I can see it on his face.*

REDUNDANCY

Redundant is a term that means "more than enough" or "overabundant." Redundancy is a serious problem in technical writing. Rewriting something twice (notice the redundancy—*rewriting* means written again, and *twice* means written again) adds extra words and sometimes adds confusion.

One common offense, and it is offensive, is using double negatives. Using double negatives is a habit that can easily be broken. Used effectively, two negatives, as in mathematics, make a positive. Read the following sentence with an intentional double negative:

It is not unlikely that he will be hired.

The two negatives (*not* and *un-*) mean that it *is* likely that he will be hired. This kind of wording is too easily misunderstood for routine business writing. It is better to word the sentence without either negative.

Now read a sentence with an unintentional double negative:

> I don't know nothing about it.

The two negatives (*-n't* and *nothing*) are redundant and confuse the real meaning of the sentence. A double-negative error is corrected simply by getting rid of one of the negatives.

> I don't know anything about it.

Some of the most slippery negative words are NEVER, HARDLY, RARELY, NONE, and NOTHING. We can turn them into positive words by replacing them with words such as ANY, EVER, or ONE. Remember that the *n't* at the end of contractions stands for *not*, a negative. Many times you will have several choices of how to correct a double negative, each correction adding a slightly different emphasis or meaning.

> *Example:* We don't never find the time we need.
> *Correction:* We don't ever find the time we need.
> We don't find the time we need.
> We never find the time we need.

> *Example:* He isn't hardly ever home.
> *Correction:* He isn't ever home.
> He is hardly ever home.

Another redundancy offense is using a noun followed immediately by its pronoun. For example, "my car, it won't start" says the noun (*car*) followed by the pronoun (*it*), which is also the car. Usually, this is a carryover from repetitious speech. In writing, repetition only adds noise. Eliminate noise when you proofread your writing.

Exercise 4.1 Rewrite the following sentences using standard, uncluttered English.

1. My old man packed his bags and he done gone. _____

2. I ain't never seen nothin' like it. _____

3. After the machine busted up, we drug it down to the service department. _____

4. She didn't know none of the other people on the committee. _____

5. Mr. Harrison he hasn't never gave no approval. _____

6. He be right back. _____

7. That book it ain't no help. _____

8. I done tried that method, and it messed me up. _____

44 *Clarity*

9. Our group we haven't finished none of that. _____

10. He seen that movie already and said it stunk. _____

COMPARISONS

A senior professor once stated that teaching is simply a matter of making comparisons: relating new, complex ideas to familiar ideas. We use **analogies,** comparisons using "as" or "like," frequently even in casual conversation. They add colorful imagery to otherwise abstract ideas.

Exercise 4.2 Write a completed analogy for the following phrases.

1. As quick as _____
2. As fast as _____
3. As smart as _____
4. As slow as _____
5. As sharp as _____

More than likely, the comparisons you used in the exercise above are common ones. Overused analogies are called **clichés** (pronounced *klee-shays*), and they should be avoided in any kind of writing, especially technical writing. If your completed analogy was predictable, it was a cliché.

The article you read, "Taking the Noise Out of Technical Writing," compares the written communication system to an electronics communications system, something tangible and familiar to electronics students. The concepts of noise filters, modulation, and signal-to-noise ratio are applied to technical writing in terms of careful writing, editorial monitoring, and relevance of a message to a particular audience.

Comparisons in descriptions liken unfamiliar objects to familiar ones to help define size, structure, and location. Calling the configuration of a low-pass filter an *inverted L-type, T-type,* or *π-type* makes it easy to distinguish and picture. We compare locations and shapes of parts to familiar anatomy: screws have heads, saws have teeth, and roads have shoulders. Circuits have elbows and legs. Sometimes we compare complex things to simple things, such as comparing *current* to water flowing through a pipe, and *protocol* to an introductory greeting or handshake.

The following example compares something technical (a word processor) to something familiar (an office).

> A word processor is a piece of software that enables a computer to function like an automated office. With the software, the computer works as a typewriter so that you can type your own files: memos, letters, reports, and graphics. The computer memory, commonly a disk, acts as a file cabinet from which you can retrieve files, edit them when necessary, and save files indefinitely. Acting as a secretary and a copier, the word processor controls the format, type font, and number of copies to be printed on the printer. Some word processors also contain a dictionary and thesaurus to check spelling and word choice and a mail merge function for automatic addressing, which relieves secretaries of these duties.

Notice the comparisons:

Office	Word Processor
Typewriter	Computer
Memos, letters, reports	Files
File cabinet	Computer memory (disk)
Secretary and copier	Printing function
Secretary	Dictionary, thesaurus, mail merge

Comparisons are useful for explaining and understanding. Find some examples of comparisons in your technical reading.

Exercise 4.3 Write a description of one of the following devices by comparing it to the nontechnical item in parentheses. Point out two or three similarities.

- schematic diagram (blueprint)
- capacitor (water bucket)
- troubleshooting a circuit (fixing a car problem)

Show your description to someone unfamiliar with the device. Then ask the person about the item. Usually, people understand new things better when they are compared to something familiar.

ANALOGIES

Many formal tests, including some employment tests, measure logical ability by analogies, a type of comparison that shows similarities between two pairs of words. The people being tested are asked to examine the relationship between two words, and then set up the same relationship with two other words.

Completed Example: fire:hot :: ice:cold

The example would be read, "Fire *is to* hot *as* ice *is to* cold. To be a true analogy, the exact relationship must exist between both sets of words: 1:2 :: 1:2, not 1:2 :: 2:1.

Exercise 4.4 Complete the following analogies.

1. tall:short :: thin: _____
2. round:sphere :: square: _____
3. cat:kitten :: cow: _____
4. outline:story :: schematic: _____
5. desk:chair :: piano: _____

Analogies can show *synonyms* (words that mean the same), *antonyms* (words that are opposites), *cause/effect,* or some other type of relationship. The best strategy for completing analogies is to put the first pair of words into a sentence that states the relationship. Then substitute the other pair of words in the same sentence, and select the set that best resembles the original relationship. Using the earlier example, we could say that fire makes things hot, and ice makes things cold.

Example: Puppy:dog :: (choose A or B)

    A. adult:child
    B. child:adult

46 *Clarity*

By turning the first set of words into the sentence, "A puppy grows up to be a dog," we can continue with answer B, "and a child grows up to be an adult." Answer A is illogical using the relationship stated in the first set of words.

Example: Up:down ::

   A. straight:bent
   B. eat:chew

This example sets an opposite relationship which is continued in answer A. Answer B is not an opposite relationship since it is necessary to chew while eating.

Example: Music:piano ::

   A. water:lake
   B. car:tires
   C. heat:resistor
   D. vibrations:sound

This is a trickier analogy. The original sentence must carefully state the relationship of music to piano to give direction in choosing the answer. Consider the statement, "All pianos are capable of making music, but not all music is made by pianos." We can eliminate answers A and B right away, since lakes do not make water, and tires do not make cars. The correct answer is C. Answer D is close, but consider the substituted sentence, "All sound is capable of making vibrations. . . ." Right away, you will notice that the cause and effect are reversed—that vibrations make sounds, not the other way around.

Exercise 4.5 Complete the following analogies by circling the letter of the similar pair. Write a sentence showing the relationship of the first pair of words to help you select the best answer.

1. Cold:ice ::
   Sentence _____

   A. warm:soda   C. spicy:liquid
   B. more:less   D. hot:boiling

2. Ticket:admission ::
   Sentence _____

   A. book:novel   C. theory:fact
   B. battery:power   D. resistor:current

3. Ac:dc ::
   Sentence _____

   A. cathode:anode   C. on:off
   B. neutron:electron   D. computer:printer

4. Amplifier:signal ::
   Sentence _____

   A. capacitor:current C. electrons:neutrons
   B. resistor:current   D. neutrons:electrons

5. Swim:pool ::
   Sentence _____

   A. run:walk   C. pen:write
   B. jog:path   D. silk:smooth

SPELLING: Doubling the Final Consonant

The problem of doubling or not doubling the final consonant when adding an ending, or suffix, is one of the easiest and most consistent spelling rules. The doubling rule is used only if the *suffix,* or new ending, begins with a vowel (ING, ED, ANCE).

First, remember that the five **vowels** are A, E, I, O, and U (we do not consider Y for this rule). All the rest of the letters in the alphabet are called **consonants.**

_____ **RULE** _____

If the last two letters in a word are a single vowel followed by a single consonant, double the final consonant before adding the ending.

We'll call this the **one and one rule.** Notice that each of the following words ends in one single vowel followed by one single consonant.

   Example: r *un* + ing = running
              h *op* + ed = hopped
              pl *od* + ing = plodding

Notice that each of the following words does not follow the *one and one* rule.

   Example: seat + ed = seated
              test + ing = testing

Note Certain letters are never doubled, even if they follow the rule. They are W, X, and Y.

   Example: draw + ing = drawing
              say + ing = saying
              box + ed = boxed

Exercise 4.6 Add endings to the following words.

1. jam + ed _____   2. band + ing _____
3. hum + ing _____   4. link + age _____
5. trim + er _____   6. drop + ed _____
7. fix + ed _____   8. loop + ing _____
9. trip + ed _____   10. ring + ing _____

There is one more part to this rule. It concerns words with more than one **syllable**—word part. When a word has more than one syllable, one of them will be **stressed**—pronounced louder than the others. Dictionaries will write an accent mark after the loudest syllable (in the phonetic spelling).

RULE

Double the final consonant of a word with two or more syllables if it follows the previous rule (one and one), and if the stress is on the last syllable.

The following words conform to the rule above, called **one and one and last rule.**

Example: refer + al = referral
submit + ing = submitting

The following words do not conform to the *one and one and last* rule.

Example: resist + or = resistor
system + atic = systematic
relax + ing = relaxing

Exercise 4.7 Add endings to the following words.

1. transfer + ence _____
2. decay + ed _____
3. emit + er _____
4. travel + ing _____
5. display + ed _____
6. control + ed _____
7. program + able _____
8. transmit + er _____
9. limit + er _____
10. admit + ance _____

Exercise 4.8 Now check your memory by writing the rule for doubling the final consonant.

VOCABULARY: SUB/SUPER Prefixes

SUPER is a prefix that can mean *more than* or *over*.

superstar = a star more famous than most others
supervisor = person responsible for others

SUB is a prefix that means *less than* or *under*.

substandard = less than the standards
submarine = underwater craft

Exercise 4.9 Complete the words using SUPER or SUB.

1. After experiencing a setback in business, some people become _____ cautious.
2. A number printed slightly above the line is called a _____script.
3. A number printed slightly below the line is called a _____script.
4. An image placed over another image is _____imposed.
5. A set of objects that is only part of a larger set is called a _____ set.
6. To repress or put down is to _____due.
7. A first, quick look at an object or situation is called a _____ficial glance.

8. Saying something twice, repeatedly, or more than enough is _____fluous.
9. A routine used only for certain purposes, less important than the main program, is called a _____routine.
10. A _____structure is the part of a building above the foundation.

Exercise 4.10 Write a sentence using the following words as adjectives. Include a definition of the word in the sentence.

1. Subconscious

Example: Subconscious behaviors are those performed without thought or deliberation.

2. superposition theorem _____

3. supernatural _____

4. sublet (*or* subleased) _____

5. supersonic _____

6. superb _____

7. subordinate _____

8. substandard _____

9. subvocal _____

10. superlative _____

WORD WATCH: Used, Supposed

USED and SUPPOSED are the past-tense and past-participle forms of the verbs *use* and *suppose*. Because the final D is sometimes unvoiced (especially when followed by "to"—*used to, supposed to*), many writers forget to add the D to the past-tense forms.

_____ **WARNING** _____

Remember to add the final D if the verb follows a helping verb such as IS or WAS. A helping verb means that the verb is in its past-participle form.

He's suppose*d* to be at work.

The supervisor is use*d* to promptness.

The present tense of these verbs will not have a final D.

I suppose you are right.

I use many components every day.

Clarity

USED is also an adjective meaning worn or secondhand.

> She sold her used books to another student.

USE is also a noun meaning the act of using, an opportunity, or a purpose.

> He could not find a use for the last resistor.
> He had no further use for the resistor.

Other forms of USE are *user, useful, usable, useless,* and *usage.*

SUPPOSE means "imagine" or "assume." In the present tense, SUPPOSE makes the writer sound uncommitted or indecisive. Avoid indecision in technical writing.

> *Weak:* I suppose I'll start a career in electronics.
> *Strong:* I'll start a career in electronics.

Other forms of SUPPOSE are *supposedly* and *supposition* (another word for a hypothesis).

Exercise 4.11 Fill in the correct form of the word.

Use USE or USED.

1. The oscilloscope was _____ to troubleshoot the design project.
2. The team could not find a _____ for the design changes.
3. The _____ equipment needed repairs.
4. Mr. Jacobson was not _____ to delays in the schedule.
5. We are _____ to meeting our deadlines.

Use SUPPOSE or SUPPOSED.

6. He wasn't _____ to start replacing parts until he finished the troubleshooting procedures.
7. How do you _____ he arrived at the correct diagnosis?
8. The manager is _____ to notify the technicians of all design defects.
9. They are _____ to record the oscilloscope measurements.
10. If you have any questions, you're _____ to ask the project director.

Exercise 4.12 Write a sentence of your own correctly using each of the following words.

1. (USE) _____

2. (USED—verb) _____

3. (USED—adjective) _____

4. (SUPPOSE) _____

5. (SUPPOSED) _____

CHAPTER FIVE

REVIEW

READING: Stress: A Deadly Wear and Tear

CHARLES SEABROOK

You are in creeping, bumper-to-bumper rush-hour traffic on the expressway. Without warning, the car in the next lane darts in front of you.

In a fraction of a second, your muscles tense, your heart pounds, and your breath quickens as you slam on the brakes to avoid a collision and vent your anger at the other driver.

This is stress. Its physical features—tightened muscles, pounding heart and other changes in the body—are the same whether the anxiety is caused by rush-hour traffic, the death of a loved one, divorce or criticism from the boss.

Scientists now believe that constant stress—a common component of today's "upwardly mobile" society—and the changes it causes in the body are taking a tremendous toll on the health of Americans, making its victims prime candidates for an early death.

During the past few years, a number of studies have found links between a person's mental well-being and physical illness. Among the findings:

- Stress may cause an increase in the body's blood cholesterol level, a major risk in heart disease. Dutch researchers Lorenz van Doormen and K. F. Orlebeke found in a six-month study of 100 accountants that there was a 20 percent increase in blood cholesterol among the men when they were under the most stress on the job. Similar studies of military personnel, students and unemployed men show increases of cholesterol from 10 percent to 20 percent during periods of extensive stress.
- Stress may weaken the immune system and make a person more vulnerable to infection. A study

A number of studies have found links between a person's mental well-being and physical illness.

"Stress: A Deadly Wear and Tear," by Charles Seabrook, from *The Atlanta Constitution,* May 14, 1985. Reprinted by permission of The Atlanta Constitution.

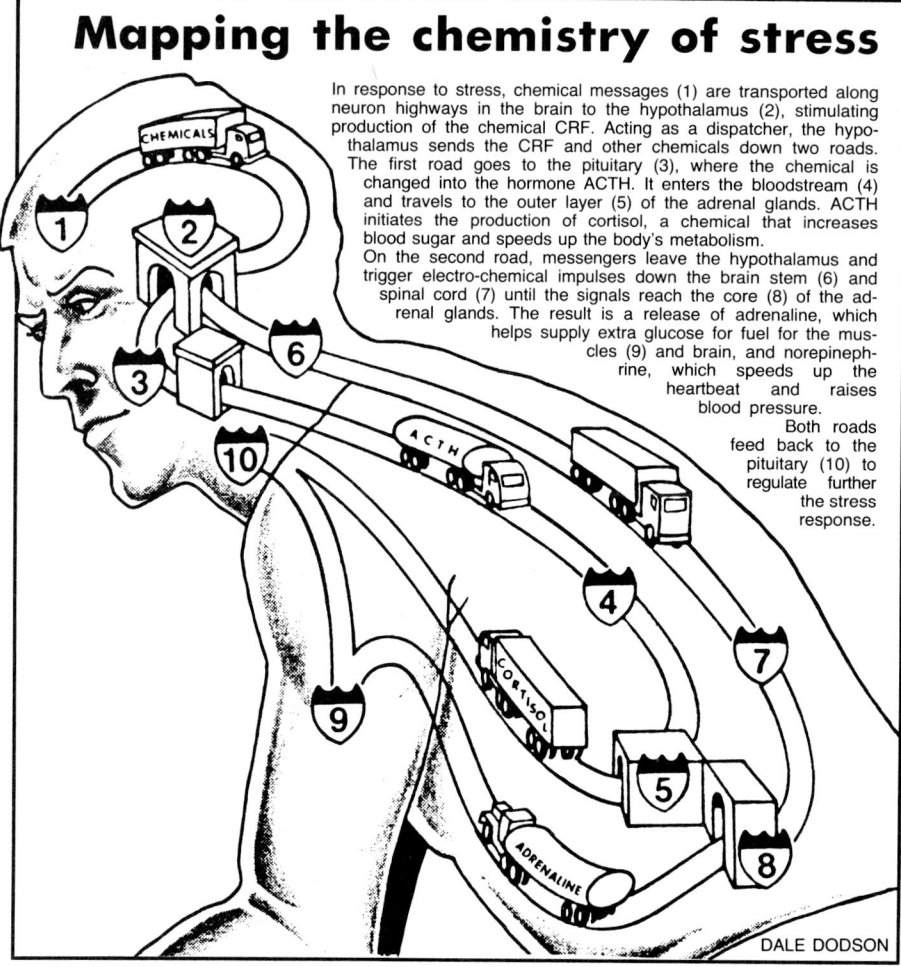

Mapping the chemistry of stress

In response to stress, chemical messages (1) are transported along neuron highways in the brain to the hypothalamus (2), stimulating production of the chemical CRF. Acting as a dispatcher, the hypothalamus sends the CRF and other chemicals down two roads. The first road goes to the pituitary (3), where the chemical is changed into the hormone ACTH. It enters the bloodstream (4) and travels to the outer layer (5) of the adrenal glands. ACTH initiates the production of cortisol, a chemical that increases blood sugar and speeds up the body's metabolism.

On the second road, messengers leave the hypothalamus and trigger electro-chemical impulses down the brain stem (6) and spinal cord (7) until the signals reach the core (8) of the adrenal glands. The result is a release of adrenaline, which helps supply extra glucose for fuel for the muscles (9) and brain, and norepinephrine, which speeds up the heartbeat and raises blood pressure.

Both roads feed back to the pituitary (10) to regulate further the stress response.

DALE DODSON

and relaxation techniques, such as biofeedback and hypnosis.

By one estimate, there are now more than 350 "stress management" enterprises in the United States. In Atlanta, one of the latest is Northside Hospital's Health Psychology Center, which focuses on the impact of mental stress on disease.

The clinic tries to identify stressful conditions that influence development of heart disease in a patient so that his chances of heart attack will be reduced.

Hospitals are turning to such clinics to help patients recover speedily so they can leave the hospital earlier. Surgery, chemotherapy and other therapies, for instance, can cause immense stress in at Mount Sinai Medical Center in New York found significant decreases in the activity of white blood cells—which help fight disease—among 16 men whose wives had died of breast cancer.

- Hostility, aggression and cynicism, which are all forms of stress, may influence the orderly ebb and flow of key hormones in the body, causing several serious health problems and perhaps even premature death. A study at Princeton University indicates that power-motivated personalities are less able to resist disease than those who maintain close relationships with other people.
- Stress-impaired immune systems may be less able to combat cancer. A study of laboratory rats at the University of Colorado suggests that people who can't cope with stress are more likely to develop cancer. Rats exposed to electric shock were given drugs that cause cells to multiply rapidly as they do in cancer. The animals' immune systems failed to produce enough white blood cells to fight the foreign substance.

"Humans are placed in a similar situation when they lose a spouse or suffer from chronic depression," says University of Colorado psychologist Dr. Steven Maier.

The new findings have spawned a number of programs in hospitals and private clinics designed to treat stress through psychotherapy

---

Possible effects of stress on:

Headache
May interfere with blood flow to the brain, resulting in painful headaches.

Infection
Lowers resistance by temporarily inhibiting some facets of the immune system.

Heart disease
May increase cholesterol levels by as much as 20 percent. A hyperactive, type A personality, and feelings of hostility also may make one more vulnerable to heart attack.

Herpes
May bring on an attack in people who have been exposed to the virus.

Ulcer
May upset the delicate flow of digestive acids in the stomach and intestines.

Stroke and kidney disease
May play a major role in causing high blood pressure, a major risk factor.

Cancer
May impair the immune system, making the body less able to ward off tumors.

patients and may inhibit their recovery from an illness.

"We offer hypnosis and biofeedback to help chemotherapy patients deal with nausea and vomiting," says psychologist Dr. Alan Lanford, head of the Northside clinic. "In many cases the patient can assert more control over the physical signs of their anxiety and handle the treatment better."

But other experts say although relaxation techniques—which appear to alter the flow of hormones in the body—have had some degree of success, more effective control of stress may not come about until the chemical basis for anxiety is fully understood.

"We are fast getting to the point where we will know exactly how stress causes chemical responses in the brain, and how these responses have unhealthy consequences for the body," says Dr. Jay Weiss of Duke University.

Both Duke and the Emory University School of Medicine in Atlanta have recently set up research programs aimed at understanding how stress, or a mental illness such as schizophrenia, influences the development of physical disease, or vice versa.

Duke researchers, for instance, are starting studies to determine how hostility and cynicism influence the health of large groups of people, including college students, rural and urban Japanese and young American adults.

Dr. Jeffrey Houpt, chairman of Emory's psychiatry department, calls the new emphasis on the relationship between mental disorders and physical disease the "repsychiatrization of medicine," in which every illness, physical as well as emotional, is considered a complex tangle of biological, emotional and social factors.

One particular type of patient that Emory's new behavioral medicine clinic will study is one who lives and behaves in such a way that he places himself at major risk for physical disease. That type includes people who smoke, drink to excess or eat until they are grossly fat, and those who purposely avoid exercise.

Reading Comprehension Questions

1. What are some physical features of stress?

2. Describe one of the links that studies have found between a person's mental well-being and physical illness.

3. What is the purpose of "stress management" enterprises? What are some of the techniques used at these clinics?

4. Describe some physical symptoms that you have experienced from stress.

5. What are some ways that you use to relieve or reduce stress?

54 Review

REVIEW: Chapters 1 to 4

Exercise 5.1 Write a summary of the article, "Stress: The Deadly Wear and Tear." Include the source, the main idea, a definition of stress, and at least five supporting details.

Exercise 5.2 Correct the 10 verb-form and subject/verb-agreement errors in the following passage.

MUSIC RELIEVES STRESS

Jogging to your favorite tunes may make exercise seem less stressful, say a researcher who found that nine runners released lower levels of a stress chemical when they listened to light rock music.

"What we have show in this simple study are already known to thousands of people who jogging while listening to music: music is help them," said an Ohio State University pharmacologist.

The study, perform with funds from Weight Watchers Foundation, have nine experienced runners jog on a treadmill for two 30-minute sessions. In the first session, five joggers wore portable stereo headphones and the others didn't. Then they switch.

Analysis of blood samples showed that the runners who listened to music had significantly lower levels of beta-endorphin, a natural opiate release by the brain in response to stress or pain. Yet music didn't caused any difference in the runners' exercising heart rates.

WORD WATCH REVIEW

In Chapters 1 through 4, Word Watch has focused on several sets of misused words—words that sound similar, yet have different spellings and meanings.

a	to	wear	they're	use
an	too	where	their	used
and	two	were	there	suppose
		we're		supposed

Exercise 5.3 In the article below, cross out the 19 errors due to misused words. Write the correct words above the error.

CHANGE TYPIFIES U.S. WORKERS TODAY

Jeffrey Golden was an radio operator during World War II, an afterward he decided their was a future in all those tubes and wires. By the late 1940s, he was working on black-and-white television sets. By the 1980s, he was peering at circuitry to tiny to see without a microscope. Today he weres a white lab jacket too work in the microelectronics plant.

When Leonard Johnson graduated from high school in 1981, he went to work as a trucker's helper, delivering freight in New York City. But he heard that secretarial work was suppose to be the wave of the future, and signed up at a school wear he completed a eight-month course in word processing. At the end of the course, the school asked him to stay on as a instructor. Today, he said with a laugh, he considers himself a "light-blue collar" worker.

Today, the old lines between blue and white collar are often blurred, some experts say. The workingman is often a woman. The sounds of work can be the soft hum and plunk of advanced automation. An, increasingly, the end product of labor is not goods, but services. Behind such trends are the lives of individual workers. In they're attitudes toward there jobs, in their perceptions of the future, are glimpses of the American worker of 1990.

Johnson, polished and confident at the age of 22, indicated that change was the only constant he was use to. "It's hard to say that anyone is in 'the right field,' " he said. "Were just getting use to one machine today, an then tomorrow that machine is outdated."

As Johnson embarks on his career, Golden, at age 63, stands just too years from retirement. In federal labor statistics, Golden would be classified as a manufacturing worker. But the quiet intensity of the Freeman Electronics Systems plant in the Bronx, were Golden works, is a world away from traditional factories.

Exercise 5.4 In the passage, the author compares two workers. In a paragraph, state the author's main idea, and support the main idea with a brief summary of the two contrasted workers.

SPELLING REVIEW
In Chapter 1 you learned the spelling rules for making plurals.

1. To change most nouns from singular to plural, we add an S or ES to the end of the noun. *Do not add* 'S (except for rule 3 below).
2. a. If the noun ends with a Y, usually change the Y to I and add ES. (frequency—frequencies)
   b. If there is a vowel before the final Y, just add the S. (delay—delays)
   c. If the noun is a proper name, just add the S. (two Jerrys)
3. If the noun is an abbreviation, number, or letter, add an S *or* an apostrophe-S (several IBM PCs, many M.D.'s)
4. If the noun is a hyphenated word, add the S to the important word. (sisters-in-law, vice-presidents)
5. Some nouns have spelling changes or foreign endings in the plural form. (man—men, medium—media)
6. Words ending in IS change to ES for the plural form. (diagnosis—diagnoses)

In Chapter 2, you learned the rules for adding LY and LLY.

General Rule: Add LY to the end of an adjective to make an adverb. (fortunate—fortunately)

1. On four words, drop the final E before adding LY. (simply, duly, truly, and wholly)
2. If the word ends in BLE, drop the final E before adding LY. (terrible—terribly)
3. If the word ends in Y, change the Y to I and add LY. (easy—easily)
4. If the word ends in C, add ALLY. (critic—critically)

56 Review

In Chapter 3 you learned the rule for EI/IE pairs.

> I before E, except after C
> Or when sounded like A, as in *neighbor* or *weigh*.

In Chapter 4 you learned the rules for doubling the final consonant.

> **One and One Rule:** If a word ends in a single vowel and a single consonant, double the final letter before adding an ending that begins with a vowel. (drop—dropped)
>
> **One and One and Last Rule:** For words of two or more syllables, if the stress is on the last syllable, apply the one and one rule. (emit—emitter)

Exercise 5.5 Vocabulary and Word Watch Puzzle The following crossword puzzle uses words found in the "Vocabulary" sections of Chapters 1 to 5. The chapter in which each word was used is given in parentheses.

ACROSS

2. two choices (1)
4. opposite of please (2)
6. base 10 number system (1)
8. adjectives include a, an, _____ (1)
9. put down or control (4)
10. two tailpipes: _____ exhaust (1)
12. three-wheeled cycle (1)
13. unusual sight (3)
17. one of a kind (1)
18. less important routine (4)
20. base 8 number system (1)
21. opposite of logical (2)
24. opposite of profit (2)
25. negative prefixes mean _____ (2)
26. explosive sound caused by traveling faster than sound: _____ boom (3)

DOWN

1. one-fourth of an area (1)
2. period of 10 years (1)
3. eight-armed sea creature (1)
5. part of a set (4)
6. opposite of code (2)
7. injury on the surface: _____ wound (4)
11. 1000 volts (1)
14. five-sided figure (1)
15. co-worker with less authority (4)
16. prefix *uni* means _____ (1)
19. of one sound: in _____ (1,3)
22. look into, investigate closely (4)
23. opposite of known (2)
26. outstanding (4)

Choose from the following words:

decade	kilovolt	sonic	the
decimal	nonprofit	spectacle	tricycle
decode	not	subdue	unique
dilemma	octal	subordinate	unison
displease	octopus	subroutine	unknown
dual	one	subset	
illogical	pentagon	superb	
inspect	quadrant	superficial	

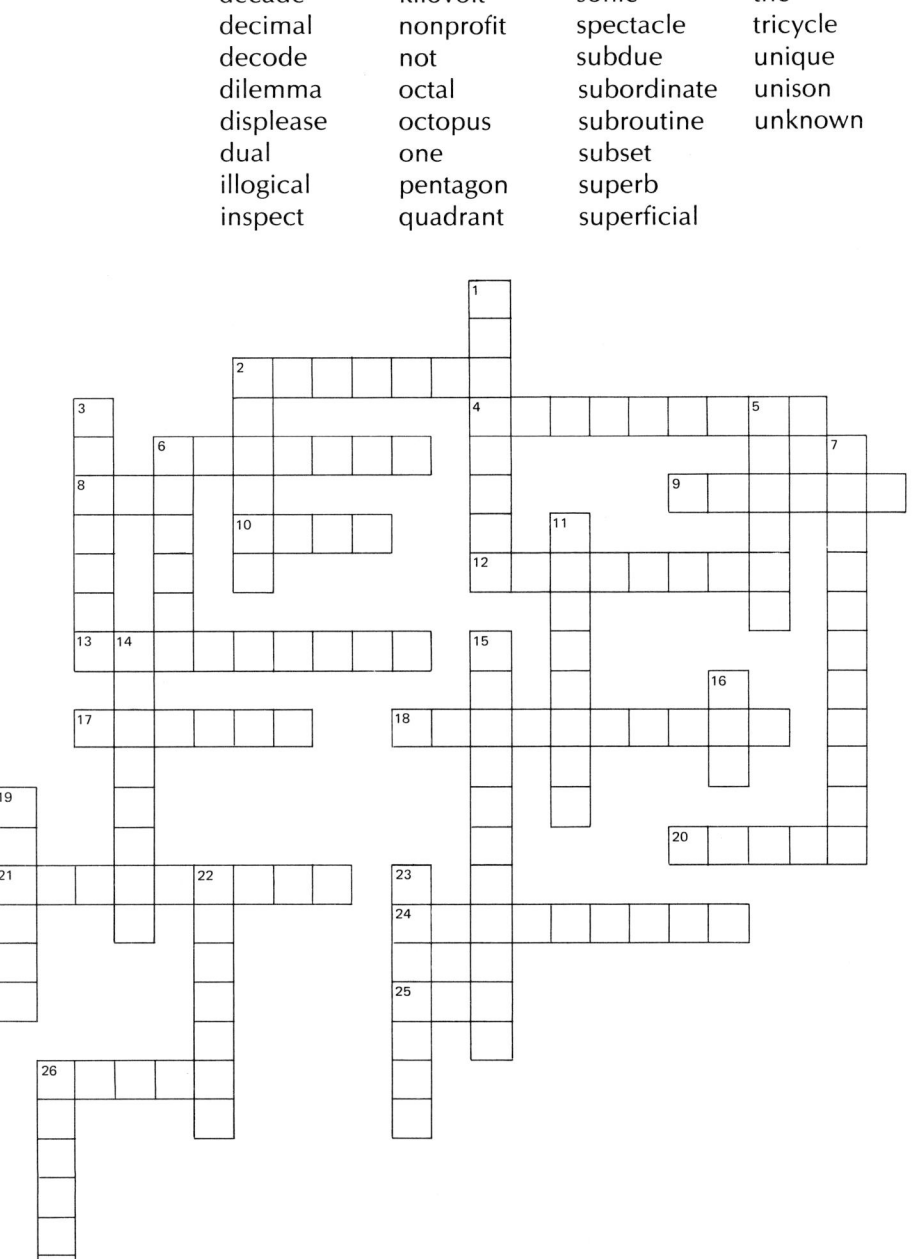

Exercise 5.6 Proofread the following passage for spelling errors. Cross out each of the 12 misspelled words and write the correct spelling above it.

LASERS LIGHTING THE WAY TO SUBMARINES

Strong beames of laser light, swept across the surface of the sea, could turn the ocean into a loudspeaker for communication with submarines', or perhaps even a new type of sonar for detectting them.

Sound already provides an important means for communicating underwater and is widely used by sea creatures such as whales, dolphins, and fish. The Navy routinelly uses sound systems (sonar) to locate submarines, and it is explorring techniques for sending messages via sound signals. Radio communication is difficult, often impossible, underwater.

The soundses generated by the laser depend on how the laser is controled. The noise can range from sharp pingging, as in sonar, to a low-frequency hum, which can be detected over long distances. Only a reciever located within a specific region will hear the sound.

It has been known for a century that a beam of light will produce heat, and the heat in turn will produce sound. The effect was proved by Alexander Graham Bell, who actualy built a device called a photophone in 1880. But the device was so impractical that it was virtually forgotten until the laser was invented in the 1950s. Research physicists are now studing applications of laser-generated sound in underwater acoustics.

REVIEW CHART FOR COMMA PLACEMENT

The following chart demonstrates comma placement in sentences. The reference tells you where to find an explanation and a practice for each situation.

Key: S: subject
V: verb
IC: independent clause
DC: dependent clause
Signal: word that subordinates (if, although, because)
Conjunction: word that coordinates (and, but, or, nor)

Situation	Model	Reference
Simple sentence	S V. (complete idea)	Grammar: Subjects and Verbs
Independent clause		
Fragment or dependent clause	Signal S V. . . .	Grammar: Fragments
Compound S, V	S and S V and V.	Grammar: Subjects and Verbs
Simple sentence with series	S V, V, and V.	Mechanics: Commas
Simple sentence with interrupting word, phrase, clause	S essential interrupter V. S, nonessential interrupter, V.	Mechanics: Commas
Simple sentence with introductory word, phrase, clause	Introductory element, S V.	Mechanics: Commas
Compound sentence with 2 ICs	S V, conjunction S V. S V; S V.	Grammar: Compound Sentences
Complex sentence with 1 IC, 1 DC	S V signal S V. Signal S V, S V.	Grammar: Complex Sentences
Compound-complex sentence	Signal S V, S V, and S V. S, signal S V, V; S V.	Grammar: Complex Sentences

Exercise 5.7 Write a sentence that demonstrates each model.

PART II

PLANNING A TECHNICAL REPORT

Chapter 6 *Comparison and Contrast*
Chapter 7 *Cause and Effect*
Chapter 8 *Review*

CHAPTER SIX

COMPARISON AND CONTRAST

- *Write a descriptive paragraph of an electronic device.*
- *Write a comparison and contrast report.*
- *Use words and figures correctly in writing numbers.*
- *Use* retro/circum/intro/intra/inter *roots correctly.*
- *Use* ough *words correctly.*

READING: Grace, Style and Intuition Contribute to Employee Success in Business

ELLIS E. CONKLIN

Executive etiquette. There is nothing scientific about it.

It's a question of grace, style, intuition—the delicate, ever-changing nuances of social behavior that, if heeded, can make the job a whole lot easier, the employer a great deal happier and the employee perhaps a little more successful.

Enter one of the masters of manners, Letitia Baldrige, chief of staff and social secretary to Jacqueline Kennedy.

"You have to know how to play the keyboard," Baldrige began in a recent interview from her home in New York City. "When you're the new kid on the block, the main thing you have to remember is to keep your distance."

Ms. Baldrige, who recently completed "Letitia Baldrige's Guide to Executive Manners" and who delivers etiquette seminars throughout the country, is referring to the eggshell relationship that can exist between employer and employee.

There are certain things you do and certain things you do not do. Ever.

"The worst thing you can do is to go to your new boss and ask him over for dinner," Ms. Baldrige said. "Let them do the inviting in the beginning. They need to make the first step."

After feeling out one's supervisor, soaking up the office atmosphere and waiting for a proper interval of time to pass, the employee's social invitation should be made privately, Ms. Baldrige added.

"And always make sure you tell the boss's secretary not to broadcast it. That's how terrible jealousies begin. If you happen to have a lucky in with the boss, don't make it known. You'll just make enemies."

"Grace, Style, and Intuition Contribute to Employee Success in Business," by Ellis E. Conklin from *The Atlanta Constitution*, April 14, 1986. Reprinted by permission of United Press International.

Judi Kaufman, a trainer for Etiquette International in Beverly Hills, said it is usually advisable that a restaurant, not the employee's home, serve as the site for the first social encounter with the boss.

"The employee, obviously, is not at the same economic level of his supervisors, so a restaurant is often a good neutral place to avoid any possible embarrassment," Ms. Kaufman said.

The old "let's-have-dinner" ritual, according to Ms. Baldrige, is only one of many social codes the eager young executive should learn and learn well.

Never call the boss by his or her first name until asked to, she cautioned. "And never should a young executive have his (or her) secretary place a call to a senior executive. There's nothing more pretentious than that."

Backslapping, making too many jokes about the job and acting too casually are the most commonly committed blunders made by ingratiating employees in the office, Ms. Baldrige stressed.

"There always has to be that distance kept," she said. "Call it respect or call it fear, but you don't do things like plopping yourself down next to the boss in the executive dining room."

In addition to serving the First Lady, Ms. Baldrige also was President Kennedy's adviser on matters of protocol.

Perhaps that's why she offered this piece of advice: "There is nothing that drives a senior executive more wild than when a junior executive barges in to the boss's office while the boss is in with someone else (to ask) questions that aren't at all urgent or important. ... That just drives them crazy."

Further, at social functions like an office party or company picnic, "Never hover around the boss and monopolize him even if you've established a friendship."

And, Ms. Kaufman added, "Don't drink too much at the party."

The best way to win over a boss, said Ms. Kaufman, is "to know his or her spouse and a little about what they're interested in, plus knowing the first names of the children."

Ms. Kaufman's list of egregious office errors include: don't take credit for someone else's work, don't be abrupt on the telephone and don't write any memos to a supervisor longer than a page.

"You'd be surprised how much an employer appreciates good telephone manners and someone who can boil all the information the boss should know into one page," Ms. Kaufman said.

Ms. Baldrige said it's the "little extra things" that breed employee success. Make sure to respond to an RSVP, she advised, and write thank you notes when someone does you a service and congratulatory notes when someone gets a promotion.

"Another thing," Ms. Baldrige continued, "we are the worst nation in the world in terms of introducing people. If you can't remember their name, laugh about it, but make the introduction."

The introductory protocol is quite simple: "If you're talking to a judge or a chief executive officer, for instance, introduce the lesser to the more important and the younger to the older.

"And when you go to a dinner party, and you're stuck between an attractive person and someone not so attractive, give each person equal attention, equal charm.

"People who are good to the little people are the ones who are going to succeed," Ms. Baldrige concluded.

Reading Comprehension Questions

1. Why is executive etiquette important?

2. What are some behaviors from the article that new employees are advised *not* to do?

3. What are some things that new employees *can* do?

4. Ms. Baldrige claims: "There always has to be that distance kept [between employees and the boss]." What do you think she meant by that statement?

5. The article concludes with the statement, "People who are good to the little people are the ones who are going to succeed." Explain this statement.

6. Describe a lesson that you have learned through your work experience that could be added to this article.

WRITING: Description, Comparison, and Contrast

Read the following excerpts from the article, a travelogue about Mexico, and a popular Travis McGee novel.

> "Once the stage is erected, the two main speaker towers are positioned on scaffolds 54 feet high. Ultimately, the Springsteen tour will present 3200 sq. ft. of loudspeakers to a single crowd. Each of the 160 speaker cabinets contains two 18-inch, low-frequency drivers; four 10-inch, lower/midrange assemblies; and two each of upper/midrange and high-frequency drivers." (Eskow, "The Heart of Rock'n Roll," *Popular Mechanics,* March 1986)

> "Inland Yucatan is a low, dense tangle of tropical jungle that gently rises from the more arid northwest to the tropical rain forest of the south. There the jungle meets and merges with Guatemala's high mountains to form the continental divide and the backbone of Central America." (Woodman, *Discovering Yucatan,* Doubleday, 1966)

> "A big man struggled out from behind the wheel and walked unsteadily to the doorway and paused there, staring at her and then at me. He was six and a half feet tall, and almost as broad as the doorway. He had a thick tangle of gray-blond hair, a mottled and puffy red face. He wore soiled khakis, with what looked like dried vomit on the front of the shirt. There was a bruise on his forehead and his knuckles were swollen. He wafted a stink of the unwashed into the small office." (MacDonald, *Under the Dreadful Lemon Sky,* Fawcett Publications, Inc., 1974)

Although the paragraphs above are quite different in their content, they have one similarity: they each communicate a picture. A descriptive paragraph has one purpose—to communicate the details of a picture in words. Providing an actual picture would make the description easier and more complete, but it is not always practical nor possible to do so.

The reading article describes business etiquette. The author provides many details: examples of good and bad office behavior. The author uses a technique called "comparison and contrast" which will be discussed later in this chapter.

The majority of technical reports include some type of description, either of objects, processes, or ideas. A description can be physical (what it looks like), functional (what it does or how it works), or both. When writing descriptions, the writer must (1) consider the technical level of the intended readers, (2) anticipate their questions, and (3) sequence the details of the description to make the picture complete and understandable.

Many professional authors labor over descriptions. They want to ensure that the descriptions will be (1) complete, (2) clear, and (3) concise.

1. **Complete** means describing all the important details of the object.
2. **Clear** means choosing the right words and the order that most accurately and logically describe the object.
3. **Concise** means using as few words as possible.

In this chapter you will write a physical description. Many beginning writers find it difficult to describe the specific details of even the most common objects: a pencil, a book cover, a room.

Certain categories of information are provided in a physical description.

| COLOR | SIZE | SHAPE |
| TEXTURE | QUALITIES | PARTS |

ARRANGEMENT OF PARTS
Each of these categories relies on vivid details that provide, in the end, a complete picture of the object. To prevent confusion, care must be taken with the order in which details are provided.

Exercise 6.1 The following description is taken from the instruction manual for a Fluke digital multimeter, model 8010A/8012A. *Underline the physical details,* those describing color, size, shape, texture, qualities, parts, arrangement of parts.

1-7. GETTING ACQUAINTED

1-8. Let's take a brief look at your instrument before we discuss exactly how to operate it.

1-9. The meter is light (2 pounds and 6 ounces for the standard model) with a low profile that hugs the work bench. The light grey case goes with any decor and is made of rugged, high impact plastic. The handle can be rotated to eight positions to function as a handle for carrying the instrument or as a stand to tilt the front panel up for convenient operation. The handle can be rotated out of the way. To change the handle position, pull out on the round hubs where the handle joins the meter then rotate the handle to the desired position.

1-10. On the rear of the meter are a phillips screw and a power cord receptacle. The phillips screw holds the outer cover in place.

1-11. The LCD (liquid crystal display) covers the left part of the front panel. The right hand portion of the front panel contains two horizontal rows of controls and connectors. The top row consists of ten pushbuttons—the four switches on the left determine the measurement function of your multimeter and the other six switches determine the range of measurement. The bottom row consists of controls and the input terminals.

TIPS FOR WRITING DESCRIPTIONS

1. Begin any description with the name of the object followed by its definition.

   Wrong: A device used to control current is a resistor.
   Correct: A resistor is a small device used to control current.

2. Describe the most obvious details first (general appearance, shape, size) and then go on to specific details (parts, arrangement). Think of how to describe the object to someone who has never seen the object.
3. Avoid abstract words such as *pretty, really,* or *very.* Such words water-down meaning.
4. Use precise, concrete adjectives or terms such as *parallel, perpendicular, cylinder,* or *grainy.*
5. Reread your description with the device in front of you, and check for completeness, logical order, and word choice.

The following physical description provides the specific details of a resistor.

> Carbon-composition resistors are small components used in electronic circuits to control current. They have two main parts: a ceramic, cylindrical casing about 3/4 inch long and two wire leads, one extending from each end. The smooth, shiny casing is light beige with three or four colored bands circling the resistor. The thin wire leads are about 1 to 2 inches long, made of tinned copper.

Exercise 6.2 Write a paragraph describing an electronic device. Include exact details. Be complete, clear, and concise. Use complete sentences. Choose one of the following devices:

- a capacitor
- a transistor
- a diode
- a chip

Show your paragraph to your instructor.

COMPARISON AND CONTRAST

When two or more objects are being compared, we often use a technique called *comparison* or *contrast.* Writers want to present facts and details in a meaningful, sometimes persuasive way, so that the reader can see the differences and similarities of the objects. The content and scope of the details will depend, naturally, on the purpose of the report: Is the writer simply informing the reader, or is the writer making an argument to persuade the reader?

The purpose of the reading article "Executive Etiquette" was clearly to persuade entry-level employees to behave in certain ways. This article could have been prepared as a list of DOs and DON'Ts. Lists of this type are effective when each item is compared and discussed in a parallel sequence.

For instance, one DO might be "Let the boss do the inviting," and the parallel DON'T would be "Ask a new boss to dinner."

Exercise 6.3 Using the article "Executive Etiquette," complete a list of DOs and DON'Ts for correct office behavior. Keep the list parallel.

Do	*Don't*
Let the boss do the inviting.	Ask a new boss to dinner.

Now compare your list to other people's lists.

WRITING COMPARISON AND CONTRAST

In a **comparison,** we look for *correlation*. In a **contrast,** we look for *differences* in certain features. It is important to determine the standards of comparison before beginning a report. Many people are familiar with the phrase "comparison shopping." When we shop, we find similar products and decide which one to purchase based on our standards, such as price, quantity, and quality. Depending on the product and how we intend to use it, one feature may be a priority. If all the products are equal in the priority feature, the decision will be made by the remaining features. For other products, we try to get the best buy for our money with no single priority.

Certain magazines, such as *Consumer Reports,* are written to compare and contrast products. Because they want their readers to trust their opinions, it is vital for them to present complete and unbiased information. Technical magazines occasionally compare and contrast new components and devices. People who are in the market for these products will read these magazines to gather pertinent information quickly. Sometimes, magazines cater to a certain audience, such as a group of people who use the products professionally. These magazines may not use cost as a feature of comparison, or may review only the most popular products, totally ignoring other worthy, but lesser-known competition. Readers need to evaluate whether an article is presenting completely neutral information or whether the magazine is biased toward a certain item or audience. As the saying goes, "Let the buyer beware!"

Comparison and contrast reports are written in three main parts: an introduction, the body of information, and a conclusion which makes a recommendation.

The first part of a comparison report, the introduction, states the standards of comparison: a short description of the features that will be discussed and why those features were chosen.

The second part of the report, the body, is the comparison and contrast of the products and their features. It is important to write parallel descriptions of each product. The features described on one product are described in the same order on all the products. If a feature is missing on a product, it is noted.

In the following chart, two pieces of test equipment are being contrasted. You already know that both the multimeter and the oscilloscope are devices used to measure ac/dc voltage, and they both have scaling adjustment knobs. Notice the criteria (features) on which the contrast will be based.

Features	*A: Multimeter*	*B: Oscilloscope*
Function	Measures resistance, ac/dc current and voltage	Measures ac/dc voltage, phase, and frequency
Display	Numerical	Graphical
Cost	$60–200	$200–800+
Complexity of use	Easy (push function buttons)	Difficult (adjust time and voltage scales)

Information in the body can be organized in two ways. The **point-by-point method** itemizes the features being examined: function, display, cost, and complexity of use in the example above. For example, if products A and B are being compared, the first paragraph might compare one feature on both products. Whenever something is said about A, the parallel information about B is presented. The next section will describe another feature of both products. The heading for each section is the feature being described.

The **block method** is organized by the products: the multimeter and the oscilloscope. First, all the features of product A are described in a block, and then the parallel features of product B are described in a block. The features of product B are described in the same order as they were described for product A. The heading for each block is the product being described.

The final section of a comparison and contrast report, the conclusion, restates only the major points that led the writer to make a final recommendation. The recommendation is a logical conclusion based on the evidence presented in the report. Sometimes the writer proposes several recommendations that take into account the possible priorities of the reader. For instance, a writer might conclude that if the reader's priority is one feature, buy product A; otherwise, buy product B. A "best buy" recommendation weighs all the features equally and makes the most cost-efficient choice.

Exercise 6.4 Write a comparison and contrast of two similar products. Pick two or more features to compare and contrast. Paragraph 1 should state the standards of comparison. Paragraph 2 should discuss one brand. Paragraph 3 should discuss the second brand. Paragraph 4 should make a recommendation to purchase one of the brands. Choose one of the following products:

- motorcycles
- cars
- compact-disc players
- computers
- modems

Share your report with other students.

SPELLING: Using Numbers

One of the effects of technology in our society is our frequent use of numbers in all types of communication. Technical writers are often unsure whether to use the word for a number or the figure for that number. The truth is that the rules are changing as fast as technology.

Style guides and textbooks differ about when and where to use figures instead of words, and some companies have their own policies about this question. Observe in your own reading how different authors and publishers treat numbers.

There are a few general rules that you can follow, although even these rules, as always, have exceptions and even alternatives.

Rule 1 Numbers that begin a sentence are ALWAYS written as words. No figures are used to begin a sentence because you cannot capitalize a figure.

> Fifty-seven dollars seemed too high for a multimeter.
> Eighty percent of the chips in that shipment were defective.
> Forty volts must be expended in overcoming the resistance.

Do not begin a sentence with a number and/or symbol that is more than four words long. Something this long is never written in words. Instead, rewrite the sentence to place the number somewhere else in the sentence.

Example

> *Wrong:* 5.9×10^{-9} of magnetic flux was detected by the magnetometer.
>
> *Right:* The magnetometer detected a magnetic flux of 5.9×10^{-9}.

Wrong: Two hundred fifty thousand dollars was awarded as a grant to the research team.
Right: A research grant of $250,000 was awarded to the team.

The following rules apply only if the numbers appear inside the sentence.

Rule 2 Numbers are written as words if either of the following situations apply:

(a) they can be written in one or two words:
He logged sixty hours of overtime.
We received 325 new training manuals.
or (b) they are below 100 (some style guides say below 10):
Only twenty of the manuals were needed.
We sent 300 manuals back.

Note. Put a hyphen (-) between two-word numbers from *twenty-one* to *ninety-nine*.

Also, if numbers over and under 100 are used in a series, all are written in figures.

Rule 3 If one number follows another number, the first is written in words and the second in figures.

three 5-ohm resistors
two 10% voltage drops

Rule 4 Precise measurements of time, distance, capacity, dimension, amount, and percent that need to be emphasized, noticed, or remembered, are written as figures. In digital notation, use the figures 1 and 0.

The horizontal scale factor is 2 ms/cm.
The largest voltage drop, 70 V, occurs across the largest resistor.
The bill included $13.50 for parts and $50 for labor.

Exercise 6.5 Rewrite the following sentences to correct any incorrect number expressions.

1. Two hundred thirty-five feet of wire were used in the prototype.

2. We completed 2 tests before we recognized the one major problem.

3. Simply multiply a decimal by one hundred to change to a percentage.

4. Connecting 2 400-V capacitors in series does not always provide eight hundred-volt capability.

5. The first compass, the lodestone, helped Chinese sailors over two thousand years ago.

Rule 5 Specific dates and addresses are written in figures. If the street name is a number, use the previous rule of numbers below 100 written in words.

>August 15, 1972
>2470 West Twenty-First Street
>452 South 152 Avenue

Rule 6 Pages, ages, and numbers of chapters, charts, and graphs are written as figures.

>Chapter 7
>Figures 4-3, 4-4
>4-year-old machine
>The company is 2 years old.

Rule 7 Fractions that express general ideas or approximations are written as separate words. If the fraction is used as an adjective (followed by a noun), it is hyphenated.

>When the tests were three-fourths completed, we stopped.
>Only two-thirds of the employees felt the stress.

Rule 8 Fractions that express exact measurements are written as figures. Mixed numbers (a whole number and a fraction) are written either with a hyphen or a space between the whole number and the fraction. The best method is the one that best clarifies your information. Be consistent.

_____ **HAZARD** _____

Remember that most electronics units use decimals or percentages rather than fractions. Decimals and most percentages are written in figures.

>$1\frac{1}{8}$-inch wire
>$1\frac{1}{8}$ inches
>2.0 kW
>80% efficiency

Exercise 6.6 Rewrite the following sentences to correct any incorrect number expressions.

1. The supervisor received 9/10ths of the credit. _____

2. The input power was fifteen point thirty-five watts. _____

3. The schematic is shown in figure 5.three. _____

4. Scott was hired on January fifth, 1980, and JoAnne was hired exactly one year later. _____

5. In the case of a twenty-μ A movement, we would need five hundred kΩ between the terminals to make the ten-volt measurement. _____

70   Comparison and Contrast

VOCABULARY: Retro/Circum/Intro/Intra/Inter

RETRO is a root word meaning "backward or behind." It is similar in meaning to the prefix RE.

> *Retrorockets* are used to slow rockets down in space.
> *Retrogression* or *regression* is the opposite of progression.

CIRCUM or CIRCU is a root word meaning "around."

> The *circumference* is the distance around.
> *Circuits* are the pathways through which current flows.

INTRO or INTRA is a root word meaning "into" or "within."

> *Intramural* teams are from within a single school.
> The *introduction* leads into the report.

INTER is a root word meaning "between" or "among."

> The scientists met to *interchange* ideas.
> The *intercom* connected all the rooms.

——————— CAUTION ———————

The prefix *in* (meaning *not*) is sometimes confused with "intro/intra." Be careful!

Exercise 6.7   Combine the correct root word to complete each sentence.

1. A pay increase that reimburses a worker for a period already worked is called _____active.
2. A/an _____venous injection goes into a vein.
3. A clever action that prevents something from happening is called a/an _____vention.
4. People who are interested only in their own minds rather than in other people are called _____verts.
5. The details surrounding an event are called _____stances.
6. A/an _____jection is a comment thrown among or interrupting a discussion.
7. An argument in which the premise is also the conclusion is called a/an _____lar argument.
8. Any unwanted signal that disturbs the reception or display of a wanted signal is called _____ference.
9. The formula for finding the _____ference of a circle is $c = \pi d$.
10. Taking a machine or device back to adapt it for a new procedure is called _____tooling.
11. An electric current that is interrupted at intervals but always flows in the same direction is called _____mittent current.
12. Paying someone back is called _____tribution.

Exercise 6.8 Combine one of the roots above with the root SPECT, meaning "look" or "see," to complete the following sentences.

A/an _____ view carefully considers all the information and circumstances before making a judgment or decision.

_____ion means analyzing one's own feelings, emotions, and behavior.

In _____ we can often find value resulting from past misfortunes.

WORD WATCH: T"OUGH" Words

One of the most troublesome letter groups in English is the OUGH group. This group of letters has unpredictable pronunciations and is found in unrelated words.

tough	ought
rough	bought
slough	sought

These words, although difficult to spell at times, have consistent pronunciations. But four words that use this letter group are trickier.

1. THROUGH is a preposition meaning "in one end and out the other" or "beyond." It is also an adjective meaning "finished." It is related to *throughout,* meaning "all the way through." *Through* is commonly abbreviated *thru* in notes, but not in formal writing. *Through* sounds the same as, but is not related to, the verb "threw," the past tense form of "throw."

    We went through the data throughout the experiment.
    We didn't stop until we were through.

2. THOUGH or ALTHOUGH are condition-setting words meaning "yet" or "in spite of."

    Even though we failed the first test, we tried again.
    We tried again although we failed the first test.

3. THOUGHT is either a past tense verb of "think" or a noun meaning "an idea." A related word is *thoughtful,* which is an adjective meaning "considerate" or "serious."

    After he thought about it, the solution seemed obvious.
    The thought of inventing a new device kept him working.

4. THOROUGH is an adjective meaning "complete" or "very exact." Related words are *thoroughbred* (of a high breed) and *thoroughfare* (a well-traveled road). The adverb form is "thoroughly."

    We checked the data thoroughly.
    The discovery was made due to his thorough research.

72 Comparison and Contrast

5. TOUGH is an adjective that usually means strong but pliant. It rhymes with "rough" (do not spell it "ruff"). The verb form is *toughen*.

   The case was made of a tough plastic.
   The high temperature toughened the ceramic.

Exercise 6.9 Using the following words, fill in the blank.

THOUGH   THOUGHT   THROUGH   THOROUGH   TOUGH

1. The researcher _____ the experiment was going well.
2. He had planned the tests _____ly.
3. Al_____ he spent several weeks planning the tests, the actual testing was finished in a few days.
4. The final test was an after _____, but it proved to be the most valuable test of all.
5. He sought the opinions of other scientists before he was _____ with his research.
6. The material used for electronic casing must be nonconductive and _____ (strong and durable).
7. The winner of the Kentucky Derby was a _____bred.
8. As Kathy walked _____ the door, she noticed that the light panel was flashing.
9. Choosing a career in electronics was not a _____ decision; nevertheless, I considered all the alternatives _____ly.
10. Everything was ready on stage before the run-_____.

Exercise 6.10 Use each word in a sentence.

1. (TOUGH) _____
2. (THROUGH) _____
3. (THOROUGH) _____
4. (THOUGH) _____
5. (THOUGHT) _____

CHAPTER SEVEN

CAUSE AND EFFECT

- *Describe solving a problem using scientific logic.*
- *Find words that signal cause-and-effect relationships.*
- *Drop the final silent e correctly when adding a suffix.*
- *Use number prefixes mono/bi/semi/poly correctly.*
- *Use effect/affect correctly.*

READING: Introduction to the Scientific Method

BARRY F. ANDERSON

Science is a body of knowledge, a social institution, or a method for acquiring knowledge, according to one's point of view. Science as a body of knowledge is treated in books on physics, chemistry, biology, psychology, sociology, and the other sciences. Science as a social institution is a topic of sociology, anthropology, and history. The concern of this book is with science as a method.

There are at least two reasons for inquiring about the scientific method. One is that it is important to understand any influence that has affected, as markedly as science has, both the shape of the world in which man lives and man's conception of his place in the world. Materially, science has, of course, added to the furnishings of the world such things as gasoline, electricity, steel, plastic, penicillin, vaccines, the automobile, the airplane, photography, radio, television, and the high-speed digital computer. Many of the effects of these material changes have been beneficial, but some, let us be honest, have been to increase man's capacity to do violence to man and to do violence to the environment on which his life depends. In the realm of ideas, science has dislodged man from his privileged place at the center of the universe (Copernicus), deprived him of much of his biological uniqueness (Darwin), and brought his rationality into serious question (Freud). At the same time, however, by ridding the world of many ghosts and by greatly extending the range of man's experience and power, science has begun to establish a new and sounder base for human dignity. Certainly, science has had, and will continue to have, a profound effect on man's life.

When one looks back into history to the time when science

"Introduction to the Scientific Method," from *The Psychology Experiment*, 2nd Edition by Barry F. Anderson. Copyright © 1966, 1966, 1971 by Wadsworth Publishing Company, Inc. Reprinted by permission of Brooks/Cole Publishing Company, Monterey, California.

showed little more promise than philosophy, the pseudosciences, folk wisdom, and magic, one wonders—much as one sometimes does in looking at a childhood picture of a great man—what was different about science that enabled it to go so much farther than its contemporaries. What were scientists doing that philosophers, astrologers, and soothsayers were not doing? Why did scientists accomplish so much more than those following other ways to knowledge? The educated man in modern society should have an answer to such questions.

A second reason for studying the scientific method is that we are all scientists in a common-sense way, and a knowledge of the methods of the formal discipline of science might help us in our everyday thinking. We all strive to build up a working knowledge of the world which will enable us to get about in it successfully. We learn what belongs in this category and what belongs in that, what to seek and what to avoid, where certain things are located, how to make certain things happen, and how to keep others from happening. We develop "theories" about the world, represented in a patchwork of likes and dislikes, proverbs, rules of thumb, analogies, and prejudices. These "theories" serve us more or less well, but the more we can improve upon them, the more effectively we should be able to conduct our lives; and it is not unreasonable to expect that a familiarity with the scientific method, the most successful method for learning about the world, might have a beneficial effect on our ability to understand the world we find ourselves in.

Indeed, the scientific method—along with exposition, logic, and mathematics—may be regarded as an extension and refinement of everyday thinking. All of these disciplines are concerned not with facts but with ways of dealing with facts; and books in these areas are all in a sense "how to" books on thinking. The great value of these disciplines lies in the fact that they are relevant to so many areas of knowledge, and the particular value of the scientific method would seem to be that it is the one most relevant to the kind of thinking we do every day, in discussing people and issues, in making practical decisions, in interpreting the news media. From that point of view, this is a book about the improvement of one's day-to-day method of acquiring knowledge about the world. It presents a particular method for acquiring such knowledge—one that has been developed over several centuries by some of history's finest minds, and one that has proved to be singularly effective.

Reading Comprehension Questions

1. Define *science*.

2. What are the two reasons given by the author for studying the scientific method?

3. What is a rule of thumb? Give an example of a rule of thumb and explain what it means.

4. The author uses the phrase "the discipline of science." What does *discipline* mean in this phrase?

5. How can using the scientific method improve your day-to-day method of acquiring knowledge about the world?

WRITING: Cause and Effect/Formal Lab Reports

THE SCIENTIFIC METHOD

To make research meaningful and understandable, scientists of all areas use the same general way of thinking about problems and solving them. These rules or steps have been used by scientists for hundreds of years. The human mind may not follow the steps exactly. Many scientists are unable to identify the stages as they are working through them. However, after the problem is solved, the scientist can use the scientific method to explain the problem and its solution in an orderly, systematic way. There are five common checkpoints.

1. *Stating the problem.* In any kind of problem solving, this is a critical stage. Identifying the problem area can determine the direction of future tests. Sometimes, superficial tests are completed to isolate the problem area and rule out other possible problems.

2. *Forming the hypothesis.* The hypothesis, or possible explanation becomes easier with experience and professional judgment. This step also relies on intuition and creativity. Some brilliant discoveries have been hypothesized after an unexplainable insight.

3. *Observing and experimenting.* By staging tests and experiments, a person can observe and record the results. The sequence of tests is described as the procedure, and the results are called data or information. When enough information is collected, the researcher goes on to the next step.

4. *Analyzing the data.* The researcher now compares and interprets the data to see what the evidence means. If the data are incomplete, further testing will have to be done (back to step 3).

5. *Drawing conclusions.* Basically, there are only two conclusions: The hypothesis was right or the hypothesis was wrong. If the hypothesis was wrong, the researcher goes back to step 2 with a revised hypothesis.

In electronics, technicians also use orderly procedures to troubleshoot, test, and maintain equipment. Rushing headlong into a series of tests wastes valuable time. Armed with a fundamental knowledge of basic electronic circuitry, technicians can functionally "divide" any electronic equipment and test it in a systematic, professional way.

FORMAL LAB REPORT

One example of the application of the scientific method to troubleshooting is in the formal lab report. A lab report documents the procedures, materials, and results of experiments. Although the exact format varies depending on the school or business setting, several standard procedures should be followed. Type the report or write it legibly in ink. Label the major divisions (except the title page) of your paper. Standard divisions include the following:

1. *Title page.* The front sheet of the report states the exact title and/or number of the experiment. It also includes your name, section, and class, the date of submission, and the instructor's name (spelled correctly).

2. *Purpose.* This section states a one-sentence objective or purpose of the experiment. While you are a student studying electronics, the objectives of lab experiments will often be given in your lab manual or by your instructor. Typical purposes include testing scientific principles, examining the effects of certain procedures on devices or circuits, and designing, modifying, and troubleshooting circuits or systems.

   Poor example:
   Objective: Common-Emitter Amplifier Troubleshooting

   Better example (list):
   Objective:
   (1) Calculate, measure, and understand the normal operation of a common-emitter amplifier circuit.
   (2) Given a circuit fault, troubleshoot its abnormal operation.

   Better example (sentence):
   The purpose of this experiment is to calculate, measure, and understand the normal operation of a common-emitter amplifier circuit, and to troubleshoot a circuit fault.

3. *Theory.* State the basic formulas, theories, or assumptions that are used in the experiment. This can be written as a list of formulas or as a paragraph of background information.

   Example of list:
   Formulas: $V_c = V_{cc} - I_c R_c$ $V_b = (R_2/R_1 + R_2)V_{cc}$
   $V_e = V_b - V_{be}\ (0.7)$ $R_{in} = R_1 \times R_2/R_1 + R_2$
   Gain $= v_{out}/V_{in}$

4. *Equipment and components.* List all the equipment and components and the number of these items that you used in the experiment. Remember that one purpose of the lab report is to direct someone who is trying to duplicate your experiment.

   Poor example:
   Resistors: 130 kΩ, 100 kΩ, 75 kΩ, 20 kΩ, 5.6 kΩ

   Better example:
   Resistors:
   130 kΩ (1)
   100 kΩ (2)
   75 kΩ (2)
   20 kΩ (5)
   5.6 kΩ (5)

5. *Procedures.* Explain in detail what you did and how you did it. Include any important illustrations, such as schematics and diagrams. Again, provide enough information so that the experiment could be repeated. Since you have already finished the experiment, use past-tense verbs. The steps can be written as a list or in paragraph form.

Poor example:
> Procedures: 1. Set up circuit using different circuit values.
> 2. Calculate and measure normal operational values.
> 3. Solve all problems given in the lab manual, utilizing troubleshooting skills.

This example sounds like a slightly reworded lab manual. It uses present-tense verbs (set, calculate, solve) and directs the reader rather than reports the experiment.

Better example:
> Procedure: 1. I constructed the circuit (see Figure 1) using the specified circuit values listed in Table 1.
> 2. I calculated and measured the normal operational values (see Table 2).
> 3. I used troubleshooting methods to solve the given problems (see Table 3).

Paragraph example:
> I constructed the circuit shown in Figure 1. Using the given circuit values in Table 1, I calculated and then measured the operational values as shown in Table 2. Finally, I solved the given problems by troubleshooting the circuits as shown in Table 3.

These two examples report the steps by using the active voice ("I" statements) and past-tense verbs. They also refer to specific figures and tables of information included in the experiment.

6. *Results.* List the raw data or results of the experiment. If charts, tables, or graphs are appropriate, draw them neatly, using a template or ruler. Label each set of information as a figure (circuit drawings or graphs) or table (lists or charts), and number them sequentially throughout the report, starting with Figure 1 and Table 1.

7. *Conclusions.* Discuss the results of the experiment, relating your results to the objective. Analyze the experiment in terms of how successfully you achieved the objective. State any points of doubt or error. Suggest modifications in the procedure that you think would improve the results. A separate discussion would state what was learned or demonstrated during the experiment and after analyzing the results.

EXAMPLE OF A FORMAL LAB REPORT

The following lab report was written using a formal lab report format as specified by an instructor. It was written in paragraph form with the figures attached at the end. Remember that each instructor may give different guidelines or requirements. Check with your instructor on specific format questions.

These students have carefully proofread their report to correct grammar, spelling, or technical errors. The result is a professional-looking report that would impress any teacher or manager.

COMMUNICATIONS LAB #7
LOW-FREQUENCY HETERODYNING

by

SHARON LINDELL
REG OLSON
CARL TALBERT

Professor H. Erickson
CS-301
March 4, 1987

EQUIPMENT USED

    Oscilloscope
    Function Generator
    Digital Multimeter
    DeVry Console 80

MATERIALS USED

    LM3900 Integrated Circuit (1)
    1N914 Diode (1)
    10-kΩ Resistors (3)
    1-MΩ Resistor (1)
    2.2-MΩ Resistor (1)
    0.002-μF Capacitor (3)
    0.2-μF Capacitor (1)

PURPOSE

The purpose of this experiment was to investigate the theory and operation of low-frequency heterodyning circuits.

THEORY

When two signals, each of a different frequency are mixed together, as shown in Figure 1, the result will be the difference in frequencies of the two signals. This output signal can then be fed into a low-pass filter with a cutoff frequency that allows the desired detected signal to pass through but filters out any undesired portion of the spectrum.

PROCEDURE

The circuit of Figure 1 was constructed on a breadboard. Next, the static voltage levels at each pin of the LM3900 IC were measured. Then, two function generators were each set to a frequency of 20 kHz and an amplitude of 50 mV(p-p) and fed into the inputs of the circuit as shown in Figure 1. The frequency of input 1 was changed to 19.5 kHz and the output was observed on an oscilloscope. This output was then sketched, indicating the amplitude and frequency. Also, the amplitude was varied between 50 and 100 mV(p-p). The output for this was again observed. Next, input 2 was reset to 50 mV(p-p) and the frequency of input 1 was varied slightly above and below 20 kHz. Finally, input was disconnected from the circuit and the bandwidth of the circuit was determined by finding the frequencies (upper and lower) at which the output amplitude is 0.707 of the amplitude of the signal at 20 kHz.

RESULTS

Output Readings of
Steps 2 and 3.
$f_{out} = 666.67$ Hz
$V_{out} = 4.5$ V(p-p)

Output Reading
of Step 4.
$f_{out} = 333.33$ Hz
$V_{out} = 7$ V(p-p)

Output Readings
of Step 5.
Above 20 kHz $f_{out} = 80$ Hz
Below 20 kHz $f_{out} = 689$ Hz

Output Readings
of Step 6.
Upper Cutoff:
  $f_{out} = 66.7$ kHz
  $V_{out} = 3.2$ V(p-p)
Lower Cutoff:
  $f_{out} = 4.5$ kHz
  $V_{out} = 3.2$ V(p-p)

DISCUSSION

With input 2 set at 19.5 kHz, the output signal was the frequency difference of the two input signals. The output signal was recorded to be 4.5 V(p-p) at 666.67 Hz. Referring to the results section, it is evident that the statement made in the theory section was supported. This is substantiated by the fact that 20 kHz − 19.5 kHz is approximately equal to the 666.67-Hz output. The error can be attributed to calibration of the function generators, the fact that they were not in phase with each other, and the fact that the two input capacitors were not exactly equal in value. The output waveform can be examined in Figure 2.

When the amplitude of input 2 was changed to 100 mV(p-p), the output amplitude increased to 7 V(p-p) and the output frequency decreased to 333.33 Hz. This change resulted in a halving of the frequency and a doubling of the output voltage. For the waveform sketch, see Figure 3.

Adjusting the frequency of input 1 caused a shift from the original frequency of 666.67 Hz. From this it can be stated that the greater the deviation from ±20 kHz, the larger the output frequency. This supports the fact that the heterodyne process does indeed take the difference of the two input signals and translates the spectrum of the one that is chosen to represent the message. The frequency response, with the high and low cutoff frequencies, can be observed in Figure 4.

CONCLUSION

In this experiment, we evaluated a low-frequency heterodyning circuit that produced the difference between two input signals. From the results section, it is readily observed that varying one of the inputs affects the output of the entire circuit. This circuit is very useful in radio broadcasting, where signals are easier to modulate at one frequency but must be broadcast at another frequency. This process can be accomplished very easily by using a heterodyne circuit.

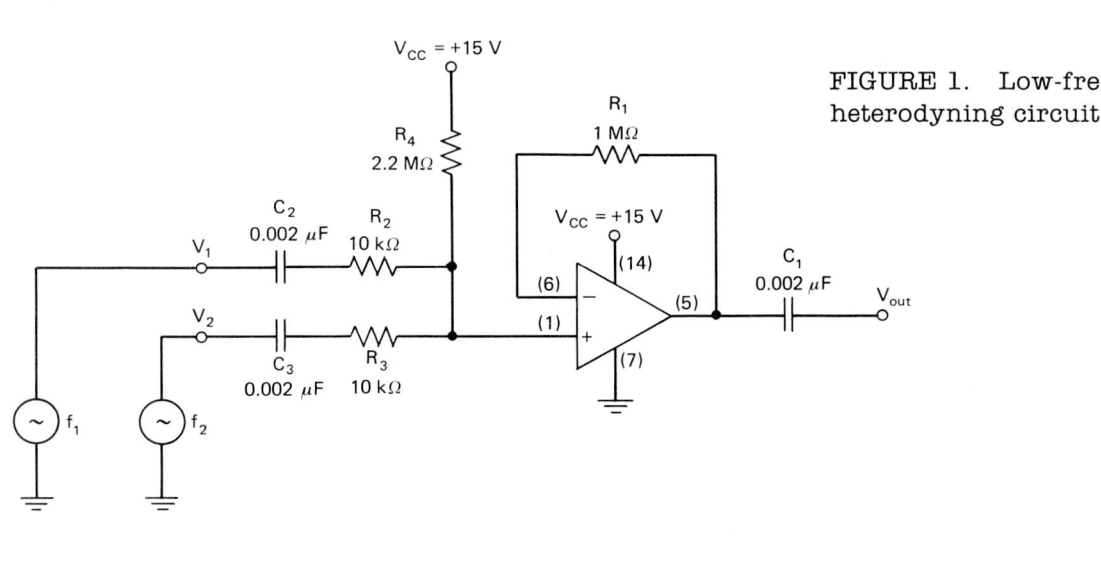

FIGURE 1. Low-frequency heterodyning circuit.

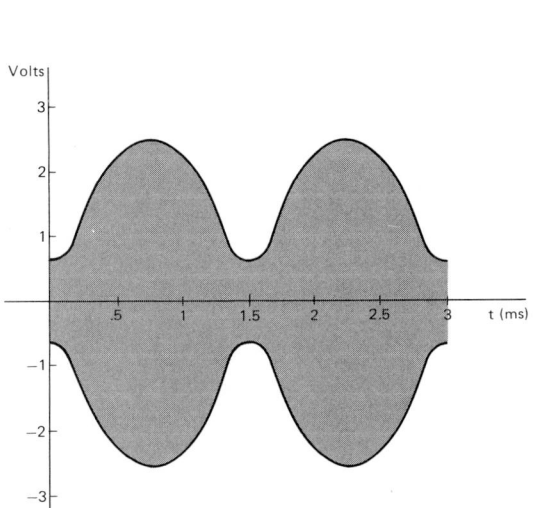

FIGURE 2. Output waveform of steps 2 and 3.

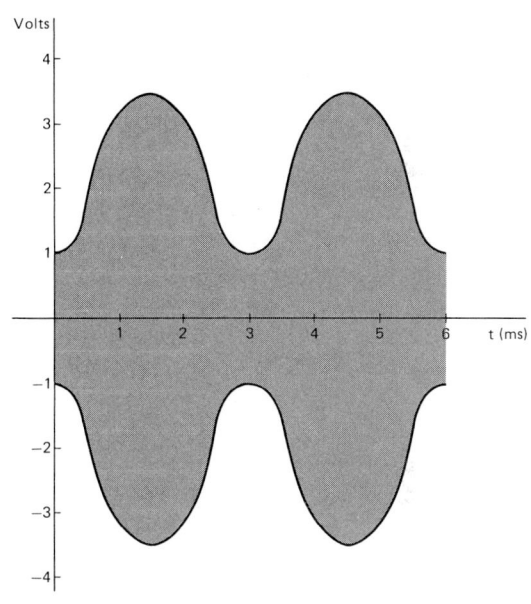

FIGURE 3. Output waveform of step 4.

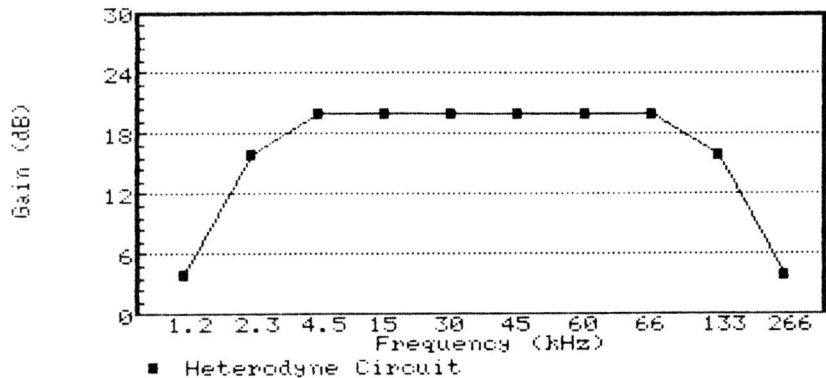

FIGURE 4. Frequency response of the output.

82 *Cause and Effect*

Exercise 7.1 *Part A.* Describe a technical or nontechnical problem that you have solved recently using scientific logic. Using the steps listed below, show how you logically arrived at your solution. An example is given below.

   Step 1. State the problem.
   Step 2. Form the hypothesis.
   Step 3. Observe and experiment.
   Step 4. Analyze the data.
   Step 5. Draw conclusions.

Example: All the food in the refrigerator is gone (*problem*).

I just bought groceries yesterday, so my roommate must have been involved last night while I was at work (*hypothesis*).

I noticed that there were lots of dishes in the dishwasher and lots of scraps in the garbage. I also noticed that glasses were sitting all over the apartment (*observe and experiment*).

One person could not have eaten all that food, dirtied all those dishes, or created that big a mess (*analysis of the data*).

From all the evidence, my roommate had a party last night (*conclusion*).

Part B. Bring in an informal lab report that you have written. Rewrite the lab as a formal lab report.

CAUSE AND EFFECT

One of the most important reading and writing skills in technical communication is the ability to identify **cause-and-effect** relationships. The entire study of science is built on this fundamental relationship. An effect is simply an observable situation that is created as the result of a cause. In other words, the cause produces the effect. A flat tire (effect) is the result of driving over a nail (cause). Most children begin to show their curiosity about the world by asking, "Why?" They have observed things (effects) they cannot understand, and they want answers (causes). Adults, too, are constantly seeking answers to problems, phenomena, and mysteries by asking, "Why?"

Many words are used to signal a cause-and-effect relationship.

therefore	consequently	thereby
yields	because	for this reason
thus	as a result	the result
hence	the cause	as a consequence
so	that is why	since

Some students use a type of shorthand notation to indicate cause and effect. Attention to the correct relationship will help the student to understand difficult concepts. Below are some examples of cause-and-effect relationships. Notice the shorthand notation of cause and effect that follows.

Example:

The series relationship among resistors has a very important *consequence* for that circuit: the current is the same at every place in the circuit.

series circuit → current same

Generally, the individual resistances in a series circuit differ from each other. *Therefore,* the individual voltages also differ from each other.

series circuit → voltage differs

Both of these ideas could be noted together in shorthand.

series circuit ⟨ current same
voltage differs

Exercise 7.2 Underline the signal word in each passage, and draw the cause/effect notation.

1. Resistance produces heat. Therefore, the temperature rises.

2. The time rate of change of current is continuously increasing in magnitude since the slope of the tangent line becomes continuously steeper.

3. As frequency increases, a capacitor's reactance becomes smaller. As a result, its ability to oppose current becomes weaker.

4. There is always the possibility of an overcurrent caused by too many loads switched ON at one time or by some malfunction in the circuit.

5. The individual voltages across the resistors in a series circuit will differ as a result of individual resistances.

SPELLING: Dropping the Final E

DROPPING THE FINAL E

In previous chapters you reviewed several rules for adding endings to words, including making words plural, changing the final Y to I, and doubling the final consonant. One situation that has not been mentioned is what to do when a word ends in a final, silent E.

use   separate
practice   compute

Rule 1 When changing the word to its S form, just add the S.

uses   separates
practices   computes

Rule 2 When adding a suffix beginning with a vowel, drop the E.

us e + ing = using
practic e + al = practical
separat e + ion = separation
comput e + er = computer

Some common exceptions are words that end with GE (pronounced J) and CE (pronounced S).

change + able = changeable
courage + ous = courageous
notice + able = noticeable
peace + ably = peaceably

Rule 3 When adding a suffix beginning with a consonant, in most cases, leave the E.

hope + ful = hopeful
love + less = loveless
move + ment = movement
separate + ly = separately

Some common exceptions are:

true + ly = truly
nine + th = ninth
wise + dom = wisdom
awe + ful = awful
argue + ment = argument
judge + ment = judgment
acknowledge + ment = acknowledgment

Exercise 7.3 Finish writing the general rules.

Rule 1. When changing a word to its S form, _____.

Rule 2. When adding a suffix beginning with a vowel, _____.

Rule 3. When adding a suffix beginning with a consonant, _____.

Exercise 7.4 List four exceptions to rules 2 and 3.

Exceptions to rule 2: _____ _____

_____ _____

Exceptions to rule 3: _____ _____

_____ _____

Exercise 7.5 Combine the root word and the suffix correctly.

1. He _____ reciting the formulas every chance he gets.
   (*practice + s*)
2. The students have become more _____ about research.
   (*knowledge + able*)
3. The entire office has made _____ progress.
   (*notice + able*)
4. The graduating students were interviewed _____.
   (*separate + ly*)
5. The resistor has become a _____ invention.
   (*use + ful*)
6. We were _____ able to contain our _____.
   (*bare + ly*)   (*excite + ment*)
7. Her _____ has improved over the term.
   (*write + ing*)
8. The invention was his own _____.
   (*create + ion*)
9. The situation seemed _____ until Dr. King took a
   (*hope + less*)
   _____ stand.
   (*courage + ous*)
10. Thomas Edison claimed that genius was "1 percent _____ and
    (*inspire + ation*)
    99 percent _____."
    (*perspire + ation*)

VOCABULARY: Other Number Prefixes (Mono/Bi/Semi/Poly)

MONO is a Greek prefix meaning single or alone. It is combined with many scientific words.

>   monochromatic = one color
>   monovalent   = having one valence

BI is a Latin prefix meaning "having two."

>   bicycle = two-wheeled vehicle
>   biceps = muscle with two points of origin as in upper arm
>   binary = made up of two parts

SEMI and its relatives, DEMI and HEMI, all mean "half" or "partially."

>   semiconductor = a partial conductor
>   hemisphere   = half the globe or sphere
>   demigod   = a minor god

Cause and Effect

POLY is a prefix meaning "many" or "more than usual." The Y is never changed to I when added to other words.

polygon = many-sided figure
polysyllabic = having several (four or more) syllables
polytechnic = providing instruction in many technical fields

Exercise 7.6 Use the correct prefix to complete each word.

1. An element with two poles is called _____polar.
2. A device with only one stable output state is _____stable.
3. A device with two stable output states is _____stable.
4. A _____conductor has electrical conductivity between that of a conductor and an insulator.
5. A lie detector test involving the measurement of many types of rate changes uses a device called a _____graph.
6. A small, half-sized cup for strong, black coffee after dinner is a _____tasse.
7. A _____tonous tone has only one pitch and can be tiresome to hear.
8. A person who knows two languages fluently is _____lingual.
9. _____styrene is a tough, colorless plastic material made by combining small molecules of styrene.
10. A job that requires only a little formal training is called _____skilled labor.

WORD WATCH: Effect/Affect

EFFECT is usually a noun meaning "the result of an action or cause." It is usually found following an adjective such as "the" or "an," and it can be made into a plural by adding S.

One of the effects of stress is irritability.
Stress can have many physical side effects.

─────────── **NOTE** ───────────

One trick for remembering this word is "Expect an Effect." Both major words begin with E. Remember that AN is a noun marker. If the use of the word is a noun, use EFFECT.

AFFECT is usually a verb meaning "to influence or produce a change." It will have an S form and also a past, a future, and a participle form.

Daily exercise affected my job performance.
The stress on this project will affect my health.
The pleasant surroundings have affected my attitude.

─────────── **NOTE** ───────────

A trick for remembering this word is "Affect is Active." Both major words begin with A. Action words are verbs. If the word is a verb, use AFFECT.

Exercise 7.7 Fill in the correct word, EFFECT or AFFECT. Add the final S or ED if necessary.

1. The supervisor had not predicted all the _____ of the tests.
2. Any type of failure can _____ a capacitor's ability to store charge.
3. Many scientific discoveries will not _____ the general society until applications are found.
4. Ohm's law states the _____ of resistance on current for a given voltage.
5. Resistance is inversely _____ by change in temperature.
6. Power can be expressed as the _____ of current and resistance.
7. The resistance of a wire is _____ by its material, length, size, and temperature.
8. One of the _____ of current flowing through a resistance is energy dissipated in the form of heat.
9. The position of resistors in a series circuit will not _____ the total resistance.
10. Increasing the conductance is one of the _____ of adding resistors in a parallel circuit.

CHAPTER EIGHT

REVIEW

READING: Coping with Stress on the Job

Stress in the workplace has been called the No. 1 health problem in the United States. In addition to its effects on individuals, its costs to American industry are estimated at $75 billion to $100 billion annually in absenteeism, diminished productivity and health-related expenses.

We also know that stress plays a major role in heart attacks, hypertension, peptic ulcers, and a host of other illnesses, from herpes to cancer.

Some sources of stress at work are so chronic and pervasive that you may accept them as a normal part of life. But your body may not acquiesce so readily. Despite an absence of outward signs of distress, a variety of harmful responses may be going on inside. Ultimately, they can lower your resistance to infection or malignancy, because of effects on the immune system or because of excessive release of chemicals and hormones that can damage your heart, stomach and other organs.

Fortunately, significant strides have been made in preventing and treating job stress. The first step is to identify the sources of stress that affect you. Some possibilities: job insecurity; lack of praise for good job performance; problems with co-workers or clients; lack of a clear job description; thwarted ambitions, or feeling that your talents are not being fully used.

While we usually associate stress with the harried, hurried middle-aged executive groping to get to the top, boring or tedious work can be equally hazardous to your health. Very often the problem is not the job itself but a mismatch between its requirements and your

PAUL J. ROSCH

Doing work you take pride in may help you live longer.

"Coping With Stress on the Job" by Paul J. Rosch, M.D. Reprinted by permission from *Nation's Business*, February 1984. Copyright 1984, U.S. Chamber of Commerce.

own goals, motivations and abilities. Job stress frequently occurs when you have a significant degree of psychological strain but have little decision-making power to influence outcomes.

Heart attacks are a major manifestation of stress. You are at greatest risk if you are a "Type A" coronary-prone individual—the hard-driving, aggressive, irritable, impatient, time-conscious person who is often frustrated by self-imposed unrealistic goals.

Type A individuals often perceive life as a series of challenges, and their bodies respond with an excessive or inappropriate release of adrenal hormones that can damage the heart.

Obviously, we cannot control some things in life—the death of a loved one, for example. Still, there are a number of ways to deal effectively with stress: removing or reducing controllable sources of stress at work; modifying behavior; using exercise as an outlet; and diminishing or blocking the harmful effects of stress by meditation or relaxation techniques.

Unfortunately, many individuals rely instead on alcohol, recreational drugs or tranquilizers to reduce stress, and these may pose other health problems.

One of the most significant recent advances has been the use of drugs known as "beta blockers," which appear to block the harmful effects of adrenalin-like hormones on the heart. ("Beta" refers to a type of receptor that receives messages from the brain—for example, to speed up the heart or to make it beat more forcefully.) In one landmark study, heart attack patients given Inderal, the original and most widely prescribed beta blocker, had a significantly lower death rate than patients not taking the drug. Similar effects have been observed for other beta blockers, such as Lopressor, Corgard and Blocadren.

Studies suggest that such drugs may also be useful in helping to change coronary-prone behavior. Type A people may actually become addicted to or dependent upon those little jolts of adrenalin to get them charged up. They may unconsciously seek ways to create that sensation by leaving things until the last minute or creating contests where none exist. By blunting the effects of adrenalin, beta blockers appear to reduce the tendency toward harmful behavior.

What is distressful for one individual may have no significance or even be pleasurable for another.

In general, doctors recognize that lacking control of a situation is distressful and can make you sick. Doing something that gives you pride, especially if you enjoy it and it benefits others, seems to be associated with good health and longevity.

Armed with that information, sit back and see how that might apply to your life. It may not be feasible to find another job, but perhaps you can identify other ways to satisfy interests and use abilities that have been suppressed. Are there ways to achieve greater control over your life by avoiding stress at work? Are there ways you can minimize harmful responses to stress?

Do not waste your energy and emotions on frustrating circumstances that you cannot possibly hope to influence. Some good advice can be found in the familiar prayer offered by the theologian Reinhold Niebuhr:

"God, give us grace to accept with serenity the things that cannot be changed, courage to change the things which should be changed and wisdom to distinguish the one from the other."

Take that philosophy to work every day and you may be a lot healthier.

Reading Comprehension Questions

1. Why is stress called "the number one health problem in the United States"?

2. What are some of the major causes of stress?

3. What are some constructive ways of dealing with stress? What are some destructive ways?

4. In general, what things do doctors consider important for good health and longevity?

5. Describe how you have coped with a stressful situation in your life.

REVIEW

The reading passage, "Coping with Stress on the Job," does not have any clearly marked transitions or headings. There are three main ideas in the passage. The points where the transitions occur have been marked with an empty box.

Exercise 8.1 Write an appropriate heading in the empty boxes for each of the three sections of the article.

Exercise 8.2 Write a descriptive essay on how you cope with stress using the following steps.

1. After reading the article, "Coping with Stress on the Job," think of all the methods you use or have used to control stress.
2. Pick two or three methods that are most effective.
3. Write an introductory paragraph listing those methods.
4. Write a paragraph on each method, describing how it helps reduce stress.
5. Proofread your paragraphs and correct any misspellings, grammatical errors, or misused words.
6. Type your final draft using a typewriter or a computer with a word processor. If you have unlimited use of a word processor, you could do steps 3 to 5 on the computer. Use the guidelines for typing in Appendix 2.

Exercise 8.3 Combine the following sentences in each group by using the first sentence as the core and eliminating any unnecessary words. You may change the order of ideas, add or delete words, or change word endings, but do not change or add to the meaning of each group.

THE HYDROGEN BOMB

Example:
    a. A hydrogen bomb is a series.
    b. The series is made up of bombs.

Combination: A hydrogen bomb is a series of bombs.

1. a. Some bombs trigger a sequence.
   b. The bombs that trigger are smaller.
   c. The sequence is reactions.

Combination: _____

2. a. The reactions result in an explosion.
   b. The reaction is called "fission."
   c. The fission is called thermonuclear.

Combination: _____

_____

3. a. The sequence begins.
   b. The beginning is a detonation.
   c. The detonation is of TNT.
   d. TNT compresses U-235.

Combination: _____

_____

4. a. The sequence causes fission.
   b. Fission releases neutrons.

Combination: _____

_____

5. a. The temperature rises.
   b. The rise is of millions of degrees.
   c. The degrees are on the Celsius scale.

Combination: _____

_____

6. a. The neutrons strike nuclei.
   b. The nuclei are lithium.
   c. The striking transforms nuclei.
   d. The transformation is into helium.
   e. The transformation is into tritium.

Combination: _____

_____

7. a. The tritium fuses with deuterium.
   b. This fusion produces more neutrons.
   c. The neutrons strike the casing.
   d. The casing is U-238.

Combination: _____

_____

8. a. This fusion produces a second stage.
   b. The stage is fission.
   c. The fission causes a release.
   d. The release is enormous.
   e. The release is of energy.

Combination: _____

_____

9. a. The release causes an explosion.
   b. The explosion has a range.
   c. The range is greater than that of the A-bomb.
   d. The range is 10 times greater.

Combination: _____

_____

10. a. The H-bomb has a "blast effect."
    b. The effect extends over 10 miles.
    c. The effect is destruction.
    d. Buildings are destroyed.
    e. The destruction is total.
    f. Or the destruction is partial.
    g. Destruction depends on nearness.
    h. The nearness is to the blast.

Combination: _____

_____

11. a. The H-bomb has a "flash effect."
    b. The effect extends over 20 miles.
    c. The effect is burning.
    d. People are burned.

Combination: _____

_____

Source: "Hydrogen Bomb" is from *Sentence Combining: A Composing Book,* by William Strong. Copyright © 1973 by Random House, Inc. Reprinted by permission of the publisher.

Exercise 8.4 Write the spelling rules that apply to the following groups of words. The chapter that reviews the rules is noted in parentheses.

1. frequency—frequencies   country—countries (1)
   Rule: _____

2. array—arrays   say—says (1)
   Rule: _____

3. knife—knives   leaf—leaves (1)
   Rule: _____

4. roof—roofs   belief—beliefs (1)
   Rule: _____

5. foot—feet   woman—women (1)
   Rule: _____

6. crisis—crises   diagnosis—diagnoses (1)
   Rule: _____

7. fortunate—fortunately   serious—seriously (2)
   Rule: _____

8. believe   conceive   eight
   relieve   receive   their (3)
   Rule: _____

9. run—running   seat—seated
   drop—dropped   test—testing (4)
   Rule: _____

10. emit—emitter   travel—traveler
    refer—referral   resist—resistor (4)
    Rule: _____

11. He logged sixty hours.
    He logged 175 hours. (6)
    Rule: _____

12. We were one-fourth finished.
    The pipe is $\frac{1}{4}$ inch in diameter. (6)
    Rule: _____

13. use—uses   move—moves (7)
    Rule: _____

14. use—using   use—useless
    move—movable   move—movement (7)
    Rule: _____

15. argue—argument   nine—ninth (7)
    Rule: _____

Exercise 8.5 Apostrophes are used to contract words and note possessives. In the following sentences, choose the correct form of the word. Refer to Mechanics Unit 2 for a review.

- If your choice is a contraction, write the two complete words.
- If your choice is a possessive, write the owner (without the 'S).
- If your choice does not use an apostrophe, leave the line blank.

Example: The (*Pentagon's*/Pentagons') newest project is called the Strategic Defense Initiative. <u>Pentagon</u>

1. The space wars envisioned by the Pentagon could be won or lost in minutes by computer programs 100 times more complex than any used in (*today's*/ *todays'*) weapons systems. _____

2. Critics of President (*Reagan's*/*Reagans'*) Strategic Defense Initiative say that it is beyond human ability to guarantee that the millions of computer instructions can be totally trustworthy. _____

3. A team of computer scientists at Georgia Tech thinks (*its*/*it's*) possible to write 10 million to 100 million error-free instructions. _____

4. (*Today's*/*Todays'*) most advanced weapons systems currently rely on only about 100,000 lines of code. _____

5. By contrast, even the computer software developed by AT&T to operate (*it's/its*) global telephone switching system contains fewer than 8 million lines. _____

6. One researcher (*says/say's*) that the SDI software has been the source of technical controversy in the computer science community. _____

7. (*They're/There/Their*) are some doubts about the feasibility of reliable programs of such complexity and accuracy. _____

8. Unless (*there's/theirs*) confidence that this system will work, there is no point in deploying it. _____

9. Pentagon (*official's/officials*) estimate that about 50 million tests may be necessary to debug the proposed programs. _____

10. A technique called "program mutation" has been developed to scatter predetermined errors throughout the code to be tested and then verify that (*they're/there/their*) are caught by the test. _____

Exercise 8.6 Complete the crossword puzzle. All the words have been taken from the "Vocabulary" and "Word Watch" sections in chapters 6 and 7.

ACROSS

3. facts of a situation (6)
5. strong, durable (6)
6. complete, careful (6)
8. prefix meaning *one* (6)
9. partial conductor (7)
10. many-sided figure (7)
11. past tense of think (6)
14. effective as of past date (6)
15. two-wheeled vehicle (7)
17. verb: to produce a change (7)
18. meaning of prefix *semi* (7)
20. noun: result of change (7)
21. two poles (7)

DOWN

1. a person interested in himself/herself (6)
2. prefix meaning backward or behind (6)
4. looking into or aware of one's own feelings (6)
7. a looking back on the past (6)
12. having a valence of 1 (7)
13. fluent in two languages (7)
16. path of electricity (6)
19. meaning of prefix *bi* (6)

Choose from the following words:

affect	half	retroactive
bicycle	introspective	retrospect
bilingual	introvert	semiconductor
bipolar	mono	thorough
circuit	monovalent	thought
circumstances	polygon	tough
effect	retro	two

Exercise 8.7 Complete the following analogies.

1. cells : skin :: atoms : _____
2. engine : run :: current : _____
3. three : nine :: four : _____
4. try : attempt :: resist : _____
5. fifteen : five :: thirty-six : _____
6. I : me :: we : _____
7. go : went :: eat : _____
8. hammer : tool :: capacitor : _____
9. sleep : rest :: eat : _____
10. opt : choose :: postpone : _____
11. superior : subordinate :: coach : _____
12. decathlon : 10 athletic events :: decade : _____
13. decode : code :: troubleshoot : _____
14. power : watts :: work : _____
15. inductor : storage :: resistor : _____

Exercise 8.8 Solve the logic problem below. It includes an introduction and a list of facts.

Clues:

1. Read the passage through once quickly to determine what type of solution is required.
2. Read the passage again, slowly, and underline the key facts.

3. Using the chart, first record the known facts by placing a cross in a box to indicate a definite IMPOSSIBILITY, and a dot to show an established FACT.
4. Use reasoning and elimination to arrive at your final solution.
5. Remember that a solution is possible using only the given facts. Every fact is important.

A young woman attending a party was introduced to four men in rather rapid succession. As usual at such gatherings, their respective types of work were mentioned rather early in the conversation. Unfortunately, she was afflicted with a somewhat faulty memory. Half an hour later, she could remember only that she had met a Mr. James, Mr. Jones, Mr. Joseph, and Mr. Johnson. She recalled that among them were a pilot, an electronics technician, an engineer, and a mechanic, but she could not recall which was which. Her hostess, a fun-loving friend, refused to refresh her memory, but offered four clues. Happily, the woman's logic was better than her memory, and she quickly paired each man with his profession. Can you?

Here are the clues:

1. Mr. Jones approached the technician for advice on fixing a TV set.
2. Mr. James had met the pilot when he hired him to fly a package to Denver.
3. The mechanic and Mr. Jones are friends, but have never had business dealings.
4. Neither Mr. Joseph nor the mechanic had ever met Mr. Johnson before that evening.

	Joseph	James	Johnson	Jones
Technician				
Engineer				
Pilot				
Mechanic				

PART

III

WRITING A TECHNICAL REPORT

Chapter 9 *Descriptive Reports*
Chapter 10 *Preparing Graphics*
Chapter 11 *Process Reports*
Chapter 12 *Review*

CHAPTER NINE

DESCRIPTIVE REPORTS

- *Write an outline for a descriptive report.*
- *Write a descriptive report based on the outline.*
- *Present an oral descriptive report.*
- *Spell words ending in ance/ence correctly.*
- *Use proto/trans/neo prefixes correctly.*
- *Use accept/except correctly.*

READING: Working with Robots: The Real Story

A few years ago we were being bombarded with predictions about the impact of robots in the workplace. Many dismal scenarios were depicted. The most dramatic envisioned the gutting of American industry by the mid-1980s. A few robot-manned, lights-out factories full of purring machinery would replace hundreds of thousands of workers who would have to spend the rest of their lives on breadlines. The labor unions could fight it tooth and nail, but the evil robot would triumph in the end.

Well, the mid-'80s are here. Did the plot fail completely, or did the robots win? Have robots robbed people of their lives, their livelihoods and their dignity?

No. Robots certainly have arrived in American industry, but without the catastrophic consequences. There have been changes—certainly a lower percentage of the work force is employed in manufacturing today than 10 years ago—but the robot's role in those changes has unfolded rather quietly, and a bit differently than most of us expected. Now that the uproar of anticipation has died down, we can look at some of the realities of life with robots.

Reality 1: Robots Aren't Replacing Entire Shifts of Workers. Many expected to see rows upon rows of robots in factories—and no people anywhere. As it has turned out, robots usually appear on the production floor only one or two at a time. And one robot doesn't replace an entire assembly line of people; it is more likely to supplant just one or two people per shift.

This is precisely what happened in the plastic molding department at Honeywell's Golden Valley, MN, plant. Troubleshooter Bob Adams reports that when the crew heard that a Puma robot was going to be installed to unload plastic molded parts, people feared for their jobs. In fact, the arrival of the robot,

CAROL FEY

The robot invasion of the American workplace was supposed to be a colossal upheaval with disastrous implications for blue-collar workers. What happened?

"Working With Robots: The Real Story," by Carol Fey. Reprinted with permission from the March 1986 issue of *Training, The Magazine of Human Resources Development.* Copyright 1986, Lakewood Publications Inc., Minneapolis, MN. All rights reserved.

99

along with other productivity improvements, saved many jobs by enabling the department to remain competitive. The robot replaced just three employees—the one person per shift who did the unloading.

Clyde Pearch, manager of training at the North American Machine Tools division of Cincinnati Milacron, speaks for many members of the robotics industry when he says that the scenario of "robots all lined up, doing what humans used to do" has not panned out. And it's not about to, either, he says. The only realistic way to look at the robot, according to Pearch, is as part of the overall manufacturing system, one component of factory automation.

Reality 2: A Robot Is Just a Tool. Sophisticated though it may be, an industrial robot is not much different from other automated manufacturing equipment.

Robotics is not necessarily new stuff, either. All of the basic knowledge and technology that goes into them has been around for years, says David Hoska, president of D&D Engineering, Inc., a Twin Cities automation consulting firm. "It's just a combination of hydraulics, mechanics, electronics and computers."

Hoska says that industry is finally beginning to look at the robot in the "right" way—as a piece of flexible manufacturing equipment. "Five years ago," Hoska reports, "I was called into plants by management and told, 'Find us a place to use a robot.'" Automated support systems for the robot usually weren't in place, and the robot often was blamed for the project's subsequent failure. A robot cannot stand alone, he points out.

Now Hoska insists on addressing automation before considering a robot. He advises against using one unless it fits into a plant's overall system.

Reality 3: Regular Plant Personnel Maintain Robots. The idea of robots as a mix of long-existing technologies is not news to the people on the factory floor who do routine maintenance on the mechanical workers. Maintenance personnel (who generally are not robotics experts—another scenario shattered) deal with the robot like any other piece of equipment. Electricians hook up the electricity, electronics and computer; plumbers bring in the air lines; millwrights move the stacking table; and the machine-repair department fixes the robot when it's broken.

Though it may seem awkward for a variety of personnel to work on every robot—and indeed it can be—there are reasons why the situation exists. One is work jurisdiction, a reality of the unionized environment. According to union contracts, each job holder has the right to do certain predetermined work. And that right is especially valued in the skilled trades—electricians, plumbers, millwrights, etc. Whenever one employee—union member or not—does another employee's work, the one whose job it was to do the work may file a grievance. If that person wins the grievance, he or she may be paid for the work, even though it has already been done.

Cubby Johnson, currently a technical trainer for robotics at a major auto assembly plant, has 35 years of experience as a union electrician. He thinks it would make sense to have an "electronic robotist," one person who could perform all the work on robots. "It may come," he says, "but we won't see it for quite a few more years. Neither the company nor the union wants to create *more* job classifications."

Many robotics experts have advanced the idea that robot-maintenance workers should be degreed specialists—graduates of vocational or university programs in robotics. But Johnson doesn't see degrees as the answer. Even without the union restrictions, the problem remains that most robotics programs offered by schools are generic. Companies still would have to train workers for specific robots.

Hoska, whose clients come from both industry and education, agrees. "The trouble with the programs we have now is there's no apprenticeship like with the trades. [Graduates] know theory but not the practical [applications]."

Even if degree programs were producing competent in-house robotics specialists, they might not be cost-effective to have on staff. "Technical specialists are very expensive," notes Nancy Johns, training coordinator at General Mills' Lodi, CA, plant. "What can you have them do during those times when their specialty isn't needed, when everything's running okay?"

Robot manufacturers tend to support the prevailing practice of training "regular" workers to handle the ordinary operation and maintenance of robots. Degreed specialists aren't necessary on a day-to-day basis, they argue. Robert Trouteaud, sales administrator for Prab Conveyers, a robot manufacturer in Kalamazoo, MI, says, "If present plant personnel can maintain programmable controllers and numerically controlled machine tools, then the addition of a robot will not be a new experience. Robot manufacturers have designed their controls to be operated and maintained by plant-level personnel."

Reality 4: Job Category and Seniority Dictate Who Is Trained. Ability, aptitude, experience and even interest have little to do with who gets selected for robotics training. Trainees usually are chosen by job category and seniority. Since both motivation and experience with robotics vary widely in such groups, it takes an unusual training program to succeed.

One of the country's most outstanding programs in robot maintenance training is at the Ford Motor Co. in Dearborn, MI. Bill Mallory, a robotics trainer for Ford, reports that even though his trainees are sent by their plant managers, "[they] are more motivated than you'd think, *if* you give them training relevant to the job. That kind of training is more useful than theory."

Ford's four-day introductory program is designed to familiarize trainees with the robots most frequently used in its plants: the Cincinnati Milacron Hydraulic T3, the Prab 4200, the Asea IRB60 and the IBM 7535. The training is practical and specific, dealing with topics like component identification, problem diagnosis and safety.

To accommodate the diversity of trainees' backgrounds, Ford's program is self-instructional and self-paced. The training takes place in a learning lab, where 24 IBM computers administer instruction to students and record their progress. Five video-disc machines and an additional 60 computers soon will expand the lab's capacity. "We handle about 45 trainees a week," says Mallory, "and we expect to double that."

In the learning lab, trainees practice manipulating a computer-controlled electric robot. This simulated experience is important, Mallory says, "because a real robot—at least the large ones used in the auto industry—is expensive and dangerous. The lab gives the trainee the first cut at programming—an opportunity to build finesse."

Once trainees acquire this finesse, they go out onto the plant floor for a "criterion performance demonstration," along with a trainer who watches closely to ensure safety.

With large robots, safety is a major issue. Despite robots' reputation for 100% predictability, they can still do things people don't expect. Mallory gives this example: "A robot can have two or three possible programs at once. If it reacts to some unanticipated change in the environment, you don't know what it will do. Or if you disable the hydraulics, the arm falls. It's not predictable to the trainee. [People] can get killed."

Bob Adams delivers a different type of on-the-job training at Honeywell. In addition to robot-maintenance people, he trains all workers in the semiskilled job of molding-machine operator. These trainees, who use robots to unload molding machines, are different from those at Ford. Since they lack the skilled-trades background that maintenance people possess, they have even less technical experience to fall back on.

For this group, Adams starts with the basics. "When I start training machine operators," says Adams, "no matter how long they've been here, I give them the benefit of the doubt by saying, 'You probably already know this but....' Then I start from ground zero. It's almost as basic as 'Here's the time clock and here's the rest room.'

"They have to understand the basic molding machine before they can understand the robot. Some of these people never received the training they should have, so they've been guessing what to do. Now they can learn what they should have known about their job all along, and still save face. No one's telling them they're dummies."

Once the trainees understand the basic molding machine, Adams explains that electrical voltages are the communication links between the molding machine and the robot. Only after trainees understand this means of communication does he proceed to how the robot operates.

Small groups of trainees gather in a classroom for the initial introduction to the basics. A slide-tape show developed by the in-house factory training department is coupled with a simple, yet technically accurate narration based on vendor troubleshooting manuals. Since Adams runs a number of classes for all three shifts, this medium offers him a painless way to deliver the same message repeatedly.

After the classroom work, the group moves to the factory floor for on-the-job practice. Adams sticks to his start-with-the-basics philosophy. To begin, trainees closely examine the molding machines they work with every day. Next they review the communication between molding machine and robot. Finally, they check out the robot itself. "I try to get them to think and use their common sense," Adams says. "I teach them the Puma by letting them realize what they already know and building on that. They learn the simple and then build to the complex. In some ways it's like using a single-cylinder lawnmower as a way to begin to understand a V-8."

The lack of relevant background tends to make the computer part of the robot the hardest for trainees to learn, he says. They've never dealt with anything like the "logic" of how a computer "thinks." "It feels strange at first to have to tell the computer when to stop as well as when to start. It would help if commands could have been designed a little easier. It takes a while to get the hang of commands like 'Do ——— mag 100.' But after a while they do learn 'computerese.'"

Once the program is ready, the trainees learn, it's just a matter of inserting the floppy disk. After they turn the power on and get memory, the computer screen tells them what to do. "Once they get used to it," Adams concludes, "there's no problem."

Frombulators

The upshot is that the reality of robots in the workplace is so mundane that it's almost disappointing. No matter what we hoped or feared robots would be, the fact is that a robot is just another tool, a piece of automation.

Fred Amram, author of a monthly column in *Robotics Today* magazine and professor of communication at the University of Minnesota, has one theory on why robots received so much attention. "By humanizing robotics we may [have been] feeding the public fear that robots will dehumanize us. ... 'What if they take over?' 'Will we become their slaves?' 'What if we can't tell a robot from a human?' 'Will they develop feelings?' "

Automation consultant David Hoska agrees that robots' unearned stigma has held back industrial installations. "The contribution of robots in industry would be higher today if they'd been called something else—'frombulator' or something—anything but 'robot.' "

The practical reality of the situation? As one assembly worker said when asked for an opinion about a newly installed industrial robot, "That thing? That's no robot—it's got no head. That's just another one of those new machines they're always bringing in here."

Reading Comprehension Questions

1. Why are companies training regular plant personnel to maintain robots rather than hiring robotics specialists?

2. Why do you think a factory worker would feel "threatened" by a robot?

3. How are robots similar to other machines? How are they different?

4. What are some advantages of choosing robotics trainees based on seniority and job category?

5. Why do you think Fred Amram, quoted in the article, suggests that we call robots some other name, such as "frombulators?"

6. What types of "automation" do you use in your day-to-day activities?

WRITING: The Descriptive Report

WRITING REPORTS

You have practiced writing sentences and paragraphs in various assignments thus far. Now it is time to put several writing skills together in the form of a report.

A report is a formal piece of writing that has a specific audience, format, and purpose. Your audience could be an instructor, a supervisor, a customer, or the board of directors. The technical level of your audience will dictate the type of information that you will include in the report. A customer, for instance, will need simply worded descriptions and explanations and more definitions of technical terms than, say, a manager who is familiar with electronics equipment, procedures, and jargon. Also, you must anticipate the needs of your audience. The board of directors may be making far-reaching decisions based on your information. They will want precise analysis and complete documentation. A customer, on the other hand, may only want to know the bottom line, such as the cost to repair a broken machine. In the remaining chapters you will be asked to write to different audiences. Take care that you write appropriately.

The format, or arrangement, of a technical report is discussed in detail in Appendix 3. Read through that section carefully if you are unfamiliar with format. Refer to it when you have specific questions and again just before you complete your final draft to double-check that you have included everything correctly. As a brief review, the standard contents are listed below.

TECHNICAL REPORT CONTENTS

　　I. PRELIMINARY SECTION
　　　　1. Title page
　　　　2. Letter of Transmittal or Preface
　　　　3. Table of Contents and List of Figures
　　　　4. Abstract

　　II. MAIN SECTION
　　　　5. Introduction
　　　　6. Body
　　　　7. Conclusion and Summary
　　　　8. Tables and Figures (if not included in body)

　　III. DOCUMENTATION
　　　　9. Quotations (footnotes or endnotes, if needed)
　　　　10. Bibliography

The purpose of the report will determine the exact format, the length, and the type of information to include. Technical reports vary in length from 1-page interoffice memos to 20-page product specifications to 200-page product proposals.

Although the amount of writing required of beginning technicians is normally limited to memos and status reports, writing responsibilities increase with experience and rank. With the growing popularity of microcomputers and word processing, many companies expect their low- and intermediate-level managers to write memos, letters, and documents without the aid of a secretary. This trend increases the need to practice good writing habits early in your career.

THE DESCRIPTIVE REPORT

A descriptive report relates the physical appearance of an object in words. It may be accompanied by pictures or line drawings that enhance the description, but pictures should never be considered a substitute for words. Most instruction manuals begin with a detailed description of the device, such as a computer or a multimeter. The description can include features, options, accessories, and specifications. You will be told what the various parts are called, how to maintain and replace them, and how to operate the device. Manuals for electronic devices also include safety information.

104 *Descriptive Reports*

In Chapter 6 you wrote a short description of an object in a short memo. As you begin writing a descriptive report, which has many paragraphs, you will quickly see the need for careful organization. Think of the report as a jigsaw puzzle. Plan the border first, and then work on units within the border. Pay attention to the relationship of one unit to another. As you will see in the example that follows, a report is arranged so that one description leads naturally to the next. Each part is described as it fits logically in the complete picture. Disjointed organization leaves the reader lost and confused.

Some basic guidelines for writing descriptions are as follows:

1. Start with an outline. Plan your report and report your plan.
2. Have the object in front of you as you write your outline. If you write about it from memory, you are more likely to omit or misrepresent important details.
3. Begin with an introduction. Lead in with general background information, a definition of the object, and a statement of the main sections of the report (a thesis sentence).
4. Use precise terms and measurements (use "toggle switch" rather than "switch," and "45 pounds" rather than "heavy"), but avoid highly technical terms and abbreviations when writing for a general audience (use "multiplex" rather than "MUX").
5. Describe the function of each part briefly, but do not confuse a descriptive report with a directions manual (discussed in Chapter 11).
6. Include a line drawing or photograph if it reinforces your description, but do not rely on the picture.
7. Label each main section of the report. Use transition and sequence words as directions for the reader.
8. Describe specific parts moving in a consistent and logical direction, such as clockwise or top to bottom, when describing a large or multifeatured object.
9. End with a summary. Restate the purpose and the thesis of the report. Opinions are not appropriate in most descriptive reports.
10. Finally, ask someone to read your description. If the reader has any questions, carefully revise the report until it is clear, concise, and complete.

EXAMPLE OF A DESCRIPTIVE REPORT

The following outline and the actual descriptive report based on the outline show an example of the logical organization of details. Since the object being described is a simple one, the report is short, and some subheadings have been combined into one paragraph. Notice, however, how the main headings of the outline are the section headings in the actual report.

OUTLINE OF THE ROBOT ARM

    I. Introduction
        A. Background
        B. Definition
        C. Divisions of the report
            1. Base
            2. Body
            3. Arm
            4. Hand

II. Base
    A. Appearance
    B. Function
III. Body
    A. Appearance
    B. Function
IV. Arm
    A. Upper arm
        1. Appearance
        2. Function
    B. Forearm
        1. Appearance
        2. Function
V. Hand
    A. Appearance
    B. Function
VI. Conclusion
    A. Restate divisions
    B. Restate purpose of robotic arm

THE ROBOTIC ARM

by

Robyn McKnight

    A technologist who can easily identify the parts of a robotic arm saves time and effort when interfacing the arm with a computer. In this report the physical characteristics of a robotic arm will be discussed in detail to aid in locating specific parts of the arm.

    A robotic arm is a mechanical device which, when interfaced with a computer, simulates the action of the human hand and arm. The major parts of the robotic arm described are the base, body, arm, and hand.

THE BASE

    The base of the robotic arm is made of blue, lightly textured metal. The base is a rectangular platform 8-1/2 inches long, 6 inches wide, and 1-1/2 inches high. The base supports the remaining parts of the arm. The body swivels relative to the base on a hollow shaft that is attached to the base.

THE BODY

    The body consists mainly of tan, lightweight plates of metal. Two main plates, each 6-1/2 by 8 inches, extend upward, housing gears and cables in the 3-inch space between the two plates. There are three motor casings on the outer side of each plate. The motors are numbered and attached to their corresponding function on the arm. The upper end of the body is connected to the arm at a shoulder joint.

THE ARM

    The arm has two parts: the upper arm and the forearm. The upper arm consists of two 10-inch plates, 2-1/2 inches apart, with drive cables housed inside the plates. An elbow connects the upper arm and the forearm.

    The forearm bends downward from the elbow. Two plates, 8-1/2 inches long and 2 inches apart, house the cables that connect the forearm to the hand at a wrist joint.

THE HAND

The orange hand is the last part of the robotic arm. It consists of two sets of links, 3 inches and 2-1/2 inches long, which function as fingers that are able to bend at a midjoint. The hand can swivel in a circular motion at the wrist joint. At the finger joint, the plates can either open or close.

The two fingers have a 4-inch spread when extended and meet when pulled together, resulting in a clamping action similar to pinching. Springs located in the hand provide the return force needed to open the hand.

CONCLUSION

All the parts of the robotic arm discussed in this report are hollow, lightweight sheets of metal. The main parts are the base, body, arm, and hand. These parts form a robotic arm which can be interfaced with a computer and follow programmed instructions.

Exercise 9.1 Write a description of a descriptive report.

Exercise 9.2 Write the missing headings for the following outline.

FUNCTION GENERATOR

I. _____
    A. Definition
    B. Casing
    C. Front-panel controls

II. _____
    A. Size
    B. Texture
    C. Color

III. _____
    A. Frequency dial
    B. Power button
    C. Frequency multiplier/mode/function controls
    D. Dc offset control
    E. Amplitude control
    F. 50 (Ω) OUT HI/LOW selectors
    G. TTL OUT connector
    H. TRIG IN connector
    I. VCG IN connector

Exercise 9.3 Write an outline for a descriptive report of one of the following objects.

- A calculator
- A circuit diagram
- A multimeter
- A wristwatch

Exercise 9.4 Write the report based on your outline in Exercise 9.3. Write it for a general audience. Include an introduction, body, and conclusion.

Exercise 9.5 Prepare an oral descriptive report that presents an introduction, body, and conclusion. Choose one of the following topics. Have the object with you during the presentation.

- A battery
- A piece of jewelry
- A favorite tie
- A car (bring a picture)

SPELLING: ANCE/ENCE Endings

The ANCE and ENCE noun-forming suffixes are used to add the meaning "an act of," "a quality or state of being," or "a thing that" to the root word.

> occurrence—the act of occurring
> resistance—the quality of being resistant
> conveyance—a means of conveying

Deciding between spelling the suffix ANCE or ENCE (similarly, ANT or ENT) can be a frustrating problem, seemingly decided long ago by someone who simply flipped a coin. There are no obvious patterns or rules for the correct ending. The unfortunate cause for this problem is that we use the Latin root word to determine the correct spelling. Those of us who have not had a formal study of Latin will have to rely on visual or artificial clues for remembering correct endings, with the exception of one minor pattern: When the verb ends in the letter R preceded by a single vowel and is accented on the last syllable, the ending is always spelled ENCE.

> pre-fer'—preference oc-cur'—occurrence
> con-fer'—conference con-cur'—concurrence
> re-fer'—reference   trans-fer'—transference

Although this rule is fairly reliable, it falls far short of the needs of most technical writers.

Some writers develop their own tricks for spelling certain words.

> attenDANCE—imagine people dancing into the room
> ambuLANCE—imagine a victim pierced by a lance
> baLANCE—imagine a ball balancing on a lance
> exisTENce—existing for ten million years
> obserVANce—imagine looking for a van

Other writers pay close attention to the words they use most frequently, and use the dictionary for lesser-used words. More words end in ENCE than in ANCE.

One bright note for electronics technicians, however, is that most electronics-related words end in ANCE.

Exercise 9.6 Write 10 electronics terms that use the ANCE ending, such as "resistance."

Check your textbooks or the dictionary if you are not sure.

1. _____ 2. _____
3. _____ 4. _____

Descriptive Reports

5. _____ 6. _____
7. _____ 8. _____
9. _____ 10. _____

Exercise 9.7 Complete the following 20 words with either ENCE or ANCE. Try them first without using a dictionary. If you are not sure, make your best guess.

1. recurr_____ 2. resist_____
3. deter_____ 4. imped_____
5. differ_____ 6. attend_____
7. observ_____ 8. disson_____
9. bal_____ 10. conduct_____
11. toler_____ 12. persist_____
13. infer_____ 14. capacit_____
15. excell_____ 16. induct_____
17. react_____ 18. exist_____
19. transfer_____ 20. reson_____

Now check your words in the dictionary.

VOCABULARY: Proto/Trans/Neo

NEO is a prefix meaning "new" or "recent."

> neo-Latin—modern Latin
> neoplasm—new, abnormal growth of tissue (tumor)

PROTO or PRO is a prefix meaning "first," "original," or "principal."

> protagonist—main character in a story
> protoplasm—essential living matter of plants and animals

TRANS is a prefix meaning "across," "over," or "through."

> transcribe—write notes over in full form
> transfer—move from one place to another

Exercise 9.8 Write the technical meaning of each word. Try to relate the meaning to the root word.

1. neo + n _____
2. neo + phyte _____
3. neo + prene _____
4. proto + col _____
5. proto + n _____
6. proto + type _____
7. trans + former _____
8. trans + ient _____
9. trans + istor _____
10. trans + mission _____

Exercise 9.9 Write the meaning of these two confusing words.

transparent _____

transluscent _____

Exercise 9.10 Using the words from Exercise 9.8, fill in the appropriate word in each sentence.

1. One of the chief parts of the atom is the _____.
2. A _____ is an amateur or a beginner.
3. Semiconductor devices used for changing ac voltage and current levels are called _____.
4. A cable or other medium over which data are sent is called a _____ line.
5. A _____ bulb, filled with a colorless gas containing two electrodes, glows when the voltage reaches firing potential.
6. The initial greeting, or "handshake," between two electronic communications devices is called _____.
7. A _____ voltage or current appears randomly without a fixed time interval between events.
8. _____ is a new, synthetic rubber that is highly resistant to heat, oil, and oxidation.
9. A transfer resistor, or _____, is a solid-state device having a collector, base, and emitter terminals.
10. The _____ is the original model of a particular type.

WORD WATCH: Accept/Except

These two words both have the same root, *cept,* which means "to take." The prefixes make them different.

> AC- means "toward" (other forms: A-, AD-, AP-, AR-, AS-)
> EX- means "from" or "away from" (a negative prefix).

Write the literal meaning of *accept:* _____

Write the literal meaning of *except:* _____

ACCEPT is a verb that means to receive or approve. The noun form is *acceptance,* and the adjective form is *acceptable.*

> The public was not ready to *accept* robots.
> Industry's *acceptance* of standards took many years.

EXCEPT is usually a condition-setting (subordinate) conjunction or a preposition meaning "take out" or "unless." The noun form is "exception." The adjective forms are "exceptionable" and "exceptional," although the second word has come to mean "outstanding."

> The robot must obey orders *except* when the orders violate the first law.
> The *exception* proves the rule.

Note EXPECT is a verb that means "to look for" or "to wait for." The root is another form of the root "spec," meaning "to see." The noun form is *expectation*.

Exercise 9.11 Fill in the correct form of *accept, except,* or *expect.*

1. A robot must obey orders given to it by a human being _____ when those orders would violate the first law.
2. A robot must protect its own existence, _____ when that would violate the first or second laws.
3. Today's robots are not sophisticated enough to _____ and obey Asimov's rules of robotics.
4. Most industrial robots do nothing _____ carry, lift, or pull things.
5. Some workers find it difficult to _____ a robot as a co-worker.
6. Modern robots use and _____ infrared and ultrasonic signals that tell when humans are around and where objects are.
7. Some people fear robots because they _____ to lose their jobs as robots invade the workplace.
8. Years ago, there were few robots around _____ for those built by hobbyists in their own basements.
9. We can _____ the robotlike toys to become more elaborate and more expensive.
10. Some programmable robots already can _____ and perform precise instructions.

CHAPTER TEN

PREPARING GRAPHICS

- *Write a functional description of a line drawing.*
- *Prepare a line drawing, and write a functional description of it.*
- *Draw a block diagram, and write a descriptive paragraph about it.*
- *Prepare a table to present information.*
- *Prepare a pie chart to present information.*
- *Prepare a bar graph and a line graph to present information, and write a paragraph explaining the graphs.*
- *Write paragraphs interpreting three self-chosen graphics.*
- *Spell words with double consonants correctly.*
- *Use tele/phono/photo and graph/gram roots correctly.*
- *Use lose/lost/loss/loose/loosen correctly.*

READING: Add Impact with Graphics

ROGER HART

A picture may be worth a thousand words but, in the business world, words rule. Traditionally, most business communication has consisted of the written word. Why? Simply because words are easier to produce than pictures or graphics.

But that is changing and changing fast. There's good reason, too.

Today, graphic representations of data and ideas are relatively easy to produce. A wide selection of new, easy-to-use computer software greatly simplifies the task of making pictures "do the talking." Too, today's personal computers are better equipped with improved graphics processors and high quality monochrome and color video display monitors. Printers, plotters and graphics cameras generate sharp, colorful visual displays to aid you in presenting difficult-to-grasp information.

Our own brain—the "computer" we use to process the volumes of business data we see each day—is composed of two parts. Our brain's left side processes and analyzes logical information, data and concepts. The right hemisphere concerns itself with spatial and artistic information and ideas. Our artistic side can often analyze complex interrelationships more effectively than if they were presented in a way that our logical, left side would have to deal with them.

Businesses are recognizing this and, as a result, using more graphic representations in communicating with co-workers and those outside the company—customers and the public. More than a million graphics software programs were pur-

"Add Impact With Graphics," by Roger Hart from *Administrative Management,* July 1986. Republished with permission from *Administrative Management,* copyright 1986, by Dalton Communications, Inc., New York.

111

chased by businesses last year. There's no doubt that graphics have arrived.

The Kinds of Graphics

There are two basic types of graphics that you can produce on your personal computer—analytical and presentation. No clear line of demarcation exists, but there are certain differences that you will recognize when examining and comparing graphics preparation software programs.

Starting with the simplest type of program, analytical graphics software allows you to create graphs and charts from data that you would otherwise show as simple rows and columns of numbers. A spreadsheet, with its many rows and columns, can give a great deal of information. But it's difficult to easily see trends and subtle changes in the data.

Analytical graphics can put these numbers into simple graphs and charts, thus making it easier to note trends and relationships.

Analytical graphics, by and large, produce simple graphics. Line graphs, bar charts, and so on, with a limited choice of text are typical. Many popular spreadsheet and financial analysis programs have the capability to generate analytical graphics from their numeric data.

The relative simplicity of analytical-style graphics makes them especially useful for presenting data and concepts to company management.

When the need is to impress, however, then presentation graphics come to the fore.

Color printing, however, exacts its toll in speed. Each color is printed separately, thus requiring its own pass over the line by the printhead. But, the impact dot-matrix printer remains a popular

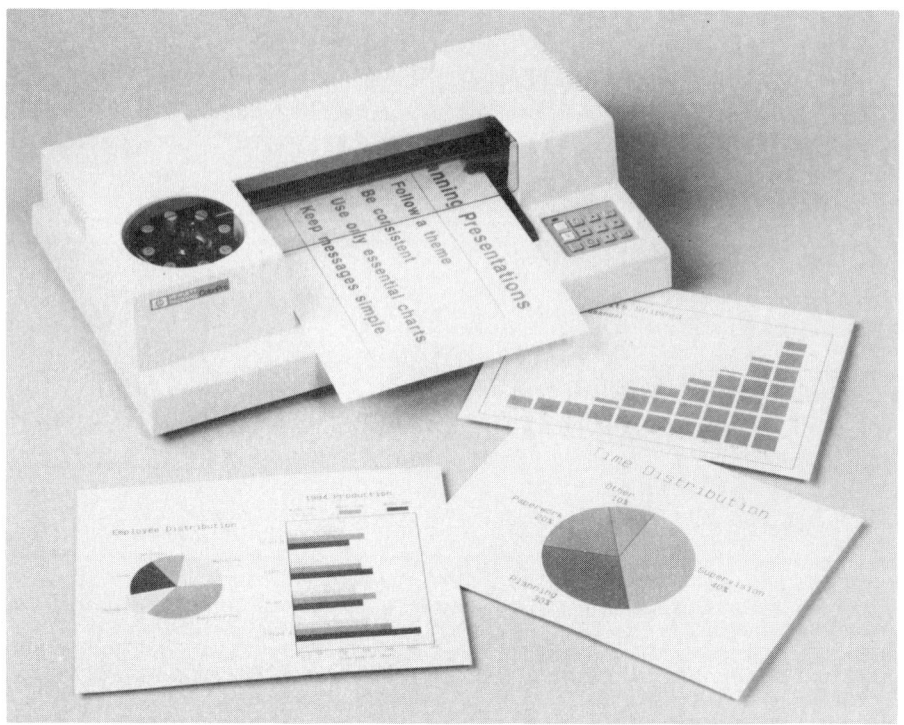

Hewlett-Packard's 7440 Colorpro business plotter combines with a personal computer and graphics software to generate high-quality text and graphics on overhead transparencies. (Photo courtesy of Hewlett-Packard Company.)

choice among graphics printing devices.

Most impact dot-matrix printers don't print well on acetate sheets. If you need to produce overhead transparencies, check the manufacturer's literature carefully and see a test print done on acetate.

Ink-Jet Printers

Another printer that is becoming more and more popular, especially for color graphics, is the ink-jet printer. Here, the ink is actually sprayed at the paper in the form of tiny droplets. This sounds messy, but the process is so well controlled that beautiful seven-color printouts emerge clean and dry.

Ink-jet printers are quiet, and the color models yield bright hues, especially suited for overhead transparencies. Early ink-jet printers were, indeed, messy, hard to add ink, and they tended to clog frequently. Today, ink-jet printers have easy-to-replace ink cartridges—no need to ever get your fingers soiled—and they are by and large self-cleaning.

If you need lots of medium-quality color graphics, especially acetate overhead slides, the color ink-jet printer is a good choice. They do form their images from dots and, like the impact models, their printing tends to be a bit coarse in appearance. This is less apparent on paper stock where the ink tends to diffuse and produce a smoother, less dot-like appearance.

Laser Printers

The superb appearance of the text characters is making the laser printer a new standard for high-quality correspondence printing. The look is close to typeset. Speed, too, is a key feature—a full page printed each 15 to 20 seconds is far beyond the reach of daisywheel printers. And, laser printers can do graphics, something that daisywheel-types can't do well, if at all.

Right now, laser printers print in black-and-white. Color models are being developed, but it's too soon to tell when they will be introduced to the market.

Like other graphics printers, laser printers make their image using dots. The difference is the large number of dots these high-tech printers use. Laser printers can put 90,000 tiny dots into a single square inch. That gives a quality to the text and graphics that has, until now, been out of reach. The less expensive laser printers use this high density dot pattern to produce text but produce graphics with a much coarser image quality. These printers are useful primarily as a replacement for a letter-quality printer.

Laser printers that also turn out super-quality graphics are more costly, but the quality is good enough to be used as camera-ready printing masters for books, newsletters, pamphlets, brochures, and magazines.

Graphics Pen Plotters

If full color graphics, in art department quality is what you're after, then you need a pen plotter. These devices work just like a graphic artist laying down straight lines, arcs, curves, circles, and lettering to achieve a finished drawing.

Plotters come in two basic types depending on how they handle the paper. Flatbed plotters hold the sheet of paper in place on a flat surface; the pens draw their lines moving in each direction to form the image. Drum plotters and roller bed plotters move the paper back and forth to draw lines in the north-south direction; the pen moves east-west. This combination creates any shape of drawing you wish, circles, arcs, etc.

Inexpensive plotters have a single pen. When the software program calls for another color to be used, you will be alerted (usually with a beep) to change pens manually. Multi-pen plotters, some with as many as eight pens, will do this automatically when called for by the software. Some will recap the pens to prevent them from drying out.

Accuracy of the plotted lines is important if the highest quality is needed. Here, look for the specifications for resolution with and without a pen change. This is a measure of how accurately the plotter will return the pen to a former position, a necessary demand if lines and arcs are to meet precisely. Lower numbers mean greater accuracy. Plotting speed can also be a buying factor if heavy use is planned.

Pen plotters can draw on paper and on acetate or polyester film for overhead transparencies. If a great many formal presentations are expected from your use of business graphics, a pen plotter will be worth the investment.

Presentation graphics programs are computerized "graphics art departments" on a disk. They allow you to create fancy charts and graphs or to take previously created analytical-style graphics and dress them up. Some programs can even produce animation and three-dimensionlike effects. The ability to add text in many sizes and typestyles, design special characters and graphic elements, like company logos, arrange and show a series of charts and graphs slide-show style are just a few of the features of many of today's presentation graphics packages.

Some programs take data from existing spreadsheet and database files to create their visual representations. Others can receive a simple analytical-style graph from your integrated spreadsheet program to allow you to use a series of graphic enhancement tools to improve their appearance.

A basic analytical graphics program may well suit your needs for in-house meetings and financial proposals. For example, customer database information can be shown as a pie chart divided into sales territory regions quickly and conveniently.

If, however, you choose to use graphics to enhance your presentations to those outside your company—customers, stockholders, or financial institutions—you will have a wide range of enhancement techniques using a presentation-grade program.

As you gain experience with the use of graphics to represent ideas, you may find yourself following a great many others in using two, three, or more packages—each for its specific set of features. Indeed, a presentation graphics program that generates text charts easily with a variety of text sizes and styles will be a boon. A large percentage of typical business presentations is in the form of word or text charts.

Getting Graphics into Your Computer

There are limitations to some computers that were not originally designed with graphics capability as a principal criterion. The IBM-PC is a good example.

Many graphics programs will require nothing more than a standard IBM-PC or compatible with 256K of memory, a standard graphics card adapter and monochrome or color monitor. Others require, or strongly recommend, that a specific brand of graphics board be used for best results. In addition, some software may need added memory or a mathematics chip microprocessor. A high-resolution color monitor will give sharp, crisp screen images that can be photographed for use as slides, if wished.

Getting images into the computer can bring added expense, too, for graphics input devices like a mouse, touch pad, light pen, drawing pad, video camera, or digitizing scanner. These first four input devices help with putting lines and other elements of drawing into the computer. The degree and accuracy of control ranges from fair with a mouse to excellent with a drawing pad.

Picture images can be put into some computers, especially those like Apple's Macintosh, which are designed with graphics in mind. Video cameras and optical scanners place three dimensional and two dimensional images into the computer for use with graphics software.

Now that you have your graphics all prepared, what do you do with them? Most people are not content to gather a crowd around the computer screen when they want to present their ideas and concepts. That means there must be a way to get the graphics out in a "hardcopy" form of a paper print or an acetate film for use as an overhead projector slide. There are actually quite a few devices to do this.

There are two basic ways that computers store graphic information. One way is to take each dot that makes up the image on the screen and translate it into a code. That's often called "bit-mapping" and it is a useful way to store and use the graphic data.

The other way to keep graphics in a computer is by means of mathematical relationships. This way, the computer remembers to store a straight line at a certain angle leading from one point to another and an arc with a certain radius from one point to a second.

Getting Graphics Out of Your Computer

Once the information is in the computer, there are three main ways to get a hardcopy of your graphic design: printers, plotters, and graphic cameras.

Printers, when they produce graphics, do it the same way newspapers produce an illustration, by breaking it into tiny dots. Every graphics printer makes its image with dots. Of course, the more dots that can be used to create a given image size, the higher quality it becomes. After all, even photographs are composed of very tiny dots of silver created by the light falling on the film. Yet, to the eye, photographic images appear solid.

Impact Dot-Matrix Printers

The standard and familiar dot-matrix printer, properly called "impact dot-matrix," makes its images by firing tiny wire pins at an inked ribbon placed over the paper. The output from these printers usually has a choppy appearance that has relegated them to doing mostly in-house presentations.

Newer designs pack a higher number of much smaller diameter wire pins into the same printing area. These high-density printers, having 18- to 24-pin printheads, turn out very sharp text and graphics.

Impact dot-matrix printers are now appearing in color, too. The older, 9-pin printhead types are becoming readily available in color models. A few of the newer high-density kinds are made in a color-printing variety, too. In addition, most of the color models will print in all-black if a solid black ribbon is installed. This easy snap-in, snap-out of ribbons gives double duty as both a standard text and color graphics printer.

To achieve color output, these printers have to shift the ribbon up and down to place the colors under the printhead. Color ribbons have horizontal bands of the three primary colors—red, blue, and yellow—to produce seven or more colors and hues.

Making Slides from Your Computer

There are two ways to make 35mm slides of your business graphics creations, too.

The simplest, cheapest method is to take a photograph of the color monitor screen. There are several kits from major film and camera manufacturers that allow the screen image to be taken with a 35mm camera. Depending on the complexity, these kits cost from just a few hundred dollars to almost two thousand.

More expensive, but of higher visual quality, are the slides produced using digital film recording cameras. Here, the unit actually creates its own high-quality image to be photographed. Sharper, clearer images can be made than relying on the quality of your monitor screen. If slides are important in your business, these high-resolution digital cameras will satisfy the most demanding needs. Costs are relatively high—$6,000 to $15,000.

The Payoff of Graphics

Of all the various communication techniques that have been studied, the use of visual graphic representations of ideas has been compared with the more traditional use of words and numbers. In every case, graphics achieved greater and faster understanding of underlying relationships and brought about agreement and consensus among the observers.

This last aspect of graphics communication—getting people to agree—is, perhaps, the true payoff of using business graphics. After all, we communicate most business ideas in order to bring about action from others, gaining management approval for a new project, cementing a relationship with a new customer, obtaining support from the financial community, rallying support for a new quality improvement program.

Businesses today are looking for better ways to communicate in a message-filled world. The use of business graphics is an important part of achieving added impact and understanding for what you have to say and ask.

Reading Comprehension Questions

1. Why are businesses starting to use more graphic representations in communications?

2. What is the difference between the two basic types of graphics, analytical and presentation?

3. What are some of the optional "input devices" that are useful for getting graphics images into the computer?

4. Describe the differences between the four main types of printers.

5. What types of graphics do you see in your textbooks? Why do you think technical books include those graphics?

WRITING: Preparing Graphics

The article "Add Impact with Graphics" is a good introduction to this vital component of technical writing. At one time, creating graphics was left in the capable hands of artists. Because of the time and expense involved, figures and drawings were infrequent and functional. With the increased capabilities of computer hardware, software, and printers, technology has broken into the graphics domain and changed it forever.

The creative and colorful nature of computer drawings makes them appealing and catchy, not only adding to our understanding and interpretation of information, but entertaining us as well. Unfortunately, graphics can be used to mislead readers, as we shall see later. Interpreting graphics correctly is as important as preparing them. In this chapter we do both.

Preparing Graphics

The main types of graphics used in technical writing are photographs, line drawings, graphs, and tables. The purpose of adding graphics to technical reports is to supplement the written material. Graphics are not used to repeat information that is already clear or to impress the reader, nor are graphics used to lengthen a report. Effective graphics can clarify information, organize data, and emphasize important points. The measures of effective graphics are simplicity and usefulness.

It is essential to plan your graphics as you outline a report. Add graphics where they are logical and useful for the reader. Explain every graphic in the text. And by all means, label each graphic with a number and a title even if your explanation is directly above or below the graphic. Add enough information in the title that a person skimming through the report would have a clear idea of the nature of the illustration—remember that some readers look at the pictures first. Normally, graphics are placed in the document just below their written explanation; however, some instructors or companies require that all graphics larger than half a page be placed in an appendix.

Each graphic, except a table, is referred to in the text as a "figure," and each is numbered starting from Figure 1 and continuing on to the last figure of the report. A table is referred to as a "table" and is also numbered starting from Table 1. Capitalize the first letter of the reference if a number follows (Table 1 or Figure 1) since it is similar to a proper noun. The examples below show a few ways to refer to figures or tables.

> A satellite dish collects signals (see Figure 1).
> The satellite dish in Fig. 1 collects signals.
> The figure shows a satellite dish that collects signals.
> Table 1 shows the values.

Use a template or ruler to draw graphics. All labels should be typed or printed neatly. Be precise but brief. If you copy your graphics from another drawing, you must footnote your drawing (use the standard footnote format shown in Appendix 3). Some original drawings credit the source of the data used in the drawing. You must also credit that source below the figure or table.

> Figure 1   Satellite dish antenna.
> *Source: Scientific Atlanta*

Because the style for each type of graphics varies, we will review some general guidelines and examples of each.

PHOTOGRAPHS

Camera-produced graphics are rarely used in technical reports because they are difficult to obtain and expensive to reproduce. They may be useful to show the overall appearance of an object, but line drawings also serve this purpose. If used, photos should always be clearly focused and include only the intended object—easier said than done. Because photographs do not necessarily indicate size, some photographs show the object next to something common, as in Figure 1. Here the reader can see the size of the satellite dish antenna in relation to a person.

LINE DRAWINGS

Line drawings include the vast majority of the graphics in your textbooks. Schematics, drawings of components, and block diagrams are all examples of line drawings. Follow these general guidelines.

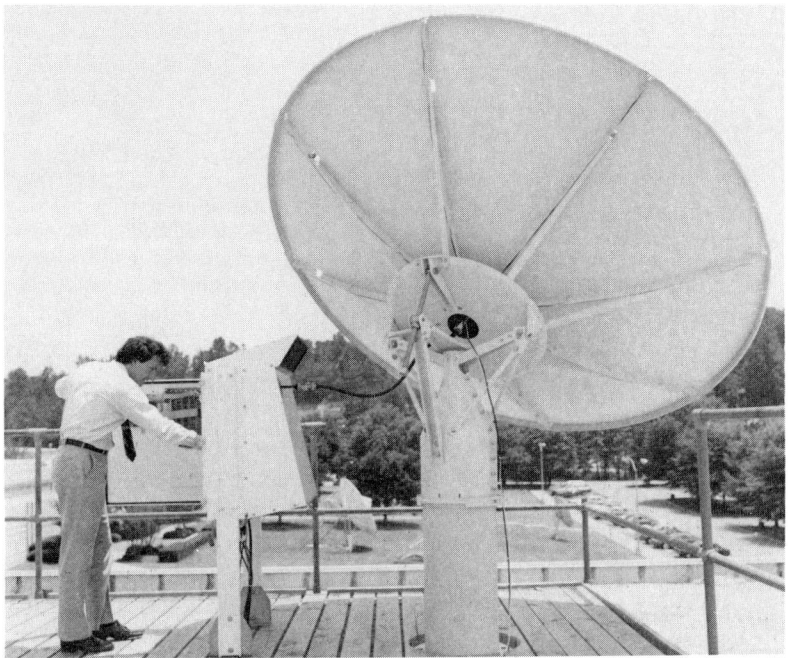

FIGURE 1 Scientific-Atlanta's IBT-1200. Ku-band transmit/receive digital earth station. Courtesy of Scientific-Atlanta.

1. Label all the significant parts of a drawing. If you use arrows or lines, they should touch the specific parts to which they point.
2. Use standard abbreviations, symbols, and terms in labels and explanations of figures. Be sure that the terms are consistent with those used in the text. Add a legend (a key for unfamiliar terms) if you are writing for a nonelectronics audience.

   V = volts   A = amperes
   Hz = hertz   ac = alternating current

3. Add enough white space so that neither the drawing nor the labels will be too crowded.
4. Use the type of line drawing that best fits your subject. These include front, side, exploded, and cutaway views, cross sections, and block diagrams.

In Figure 2 you see two diagrams of an electric circuit. The pictorial diagram (A) is labeled since it would be used for nonelectronics people. The schematic drawing (B) is unlabeled since it would be read by technical people who understand the symbols used to represent specific components.

FIGURE 2 A simple electric circuit may be represented by a pictorial diagram (A), which involves drawings of the electrical components, or by a schematic diagram (B), which consists of interconnected standard symbols used by electricians to depict specific components. Reprinted with permission of *Academic American Encyclopedia,* © 1986 by Grolier, Inc.

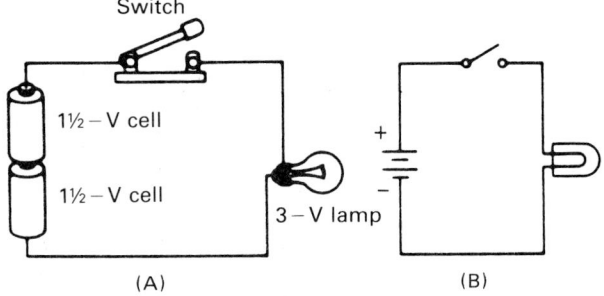

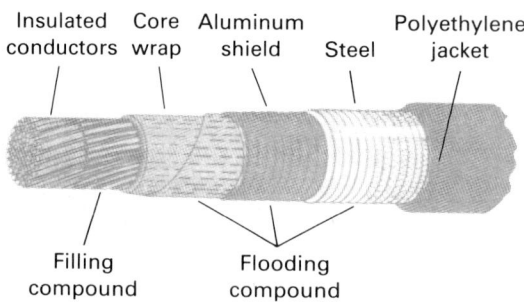

FIGURE 3 Cutaway drawing of waterproof cable design. "Communication Cables," Vol. 3, p. 442. Reprinted with permission of *McGraw-Hill Encyclopedia of Science and Technology,* © 1982 by McGraw-Hill Book Company.

In Figure 3 you see a cutaway drawing of a waterproof communications cable. The labels are close to the parts. Outer layers are cut away in the diagram to expose the inner structure.

In Figure 4 you see a process drawing of a membrane touchpad. The second stage, as the pad is touched, displays the internal change of the pad. All parts should be labeled in the first drawing; similar drawings do not necessarily have to be labeled. Consistent labels are important to show how parts change or move in a series of steps. The explanatory notes describe the process while referring to the drawing.

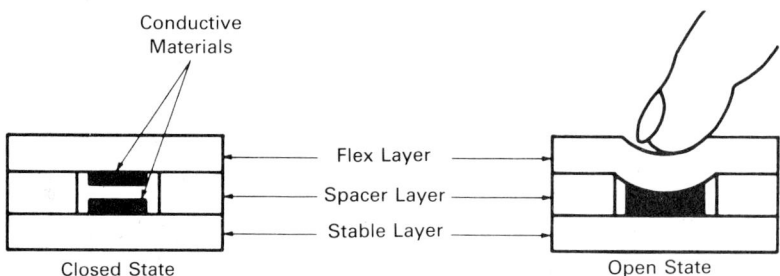

FIGURE 4 Operation of a conventional flat-panel membrane switch. Courtesy of Dupont.

In Figure 5 you see an exploded view of a satellite. Exploded views take devices apart to show internal structure and how parts fit together. The view shows 14 separate parts of the satellite. The parts are numbered and labeled. You will read more about this satellite in Chapter 11.

Figure 6 is a block diagram, or signal flowchart, of a color TV transmitter. Many electronic and mechanical processes are more simply explained by using block diagrams. Each block represents a functional unit or step, but the details of the unit are not included. The signals (lines, links, or arrows) show communications or relationships between units. Follow standard conventions when preparing these displays. Lines that cross should clearly indicate whether the lines make a connection (denoted with a darkened dot) or do not make a connection (denoted with a semicircle or arch in one of the intersecting lines).

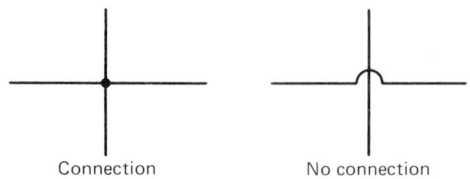

Use dashed lines inside blocks to show subunits. Provide just enough notation to make the diagram easy to understand.

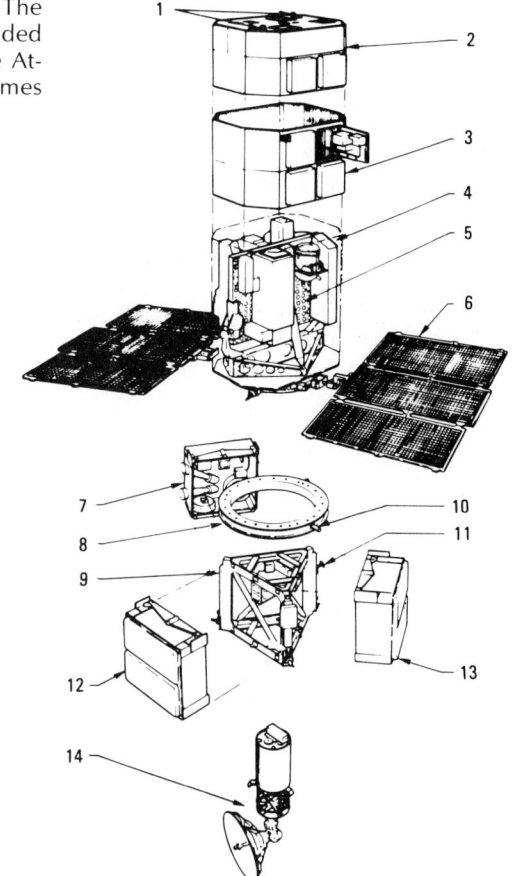

FIGURE 5 An exploded view of the Solar Max satellite. The three modular units at the base of the satellite are intended for replacement while in space. One of those units, the Attitude Control System (ACS), was replaced in space by James D. van Hoften and George D. Nelson.

1. Coarse sun sensors
2. Thermal enclosure
3. Electronic enclosure
4. Instruments
5. Instrument support plate
6. Solar array system
7. Attitude control system (ACS) module
8. Transition adapter
9. Module support structure
10. Trunnion pin
11. Grapple point
12. Power module
13. C and DH modules
14. High-gain antenna system (HGAS)

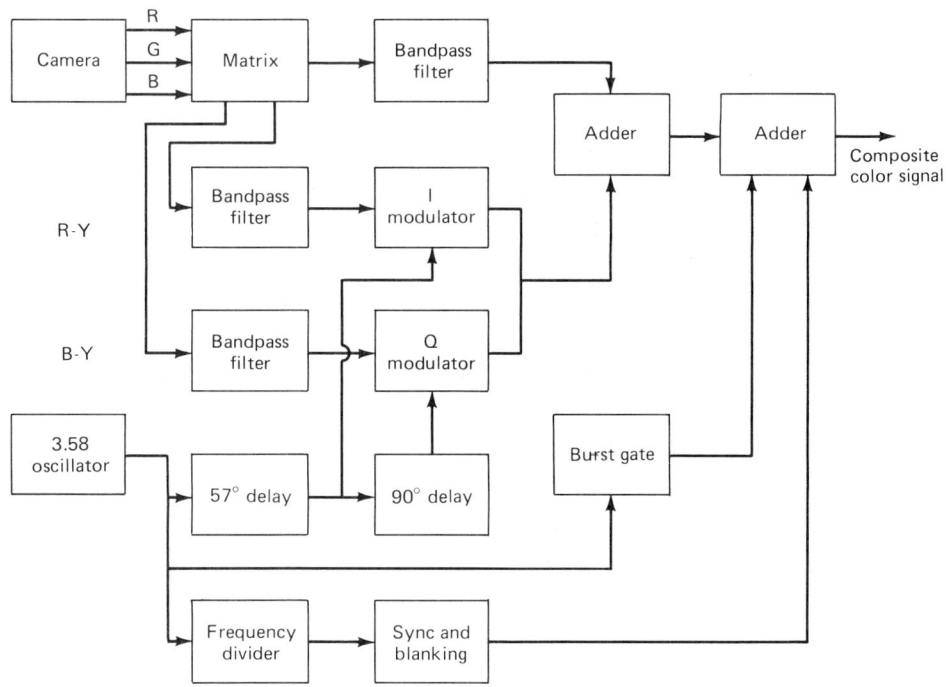

FIGURE 6 Block diagram of color TV transmitter. Courtesy of Fred Kerr.

GRAPHS

Graphs include displays of numerical data using bars, lines, curves, and circles. The purpose of graphs is to visualize the effects of a changed variable on a subject. The display often emphasizes a trend or illustrates the results of an experiment. The graphs you know best are waveforms and exponential curves.

1. The horizontal and vertical axes should be clearly labeled (frequencies, voltages, time).
2. The increments should also be marked and labeled clearly. Increments should be regular (every 10 ms, every 20 mA). The lowest value (origin) is usually zero. The maximum value is usually just one increment higher than the highest value to be represented on the graph. An arrow at the outer point of an axis represents infinite increments.
3. Graphs sometimes contain shaded or figured areas, to provide emphasis or a visual contrast to sets of numbers: the shaded side represents one set, the unshaded represents another.

Bar graphs show evenly spaced bars extending vertically or horizontally. Some writers print the exact value inside each bar, which is especially helpful when precision counts.

The bar graph in Figure 7 shows the number of cars sold from 1981 through 1985, with estimates for 1986 and 1987. The legend explains the meaning of the dark and light shadings. According to the graph, the peak in auto sales was 1985, at $11 million. The lowest point was 1982, at $8 million. The trend shows a decline in sales in 1986 and again in 1987, although neither year will decline to the sales of 1982.

Line graphs and waveforms are made up of dots placed at coordinates according to fixed increments on the vertical and horizontal axes. The dots are then connected by straight lines or smooth curves to show the subject's response to changing conditions. The horizontal axis usually represents the changed condition,

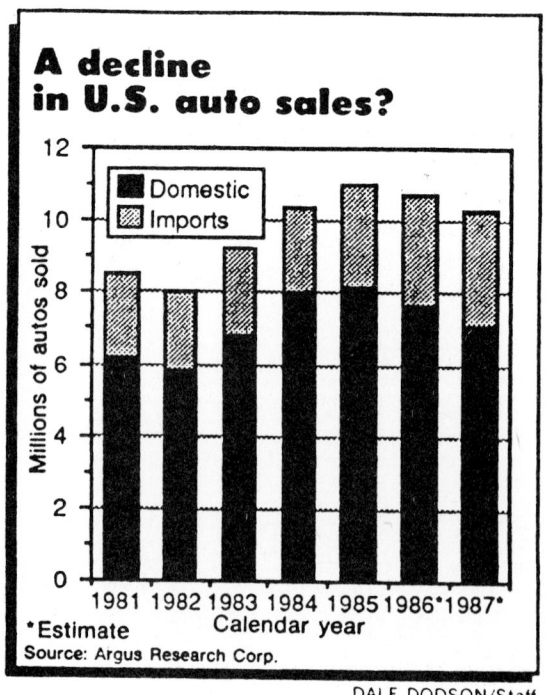

FIGURE 7 Bar graph of auto sales. Reprinted with permission, *The Atlanta Constitution*, 10/4/86.

and the vertical axis usually represents the subject's response or activity. Line graphs are particularly useful for displaying patterns and for predicting future activity. Normally, both axes begin with zero in the lower left corner, but in electronics, this is not always the case.

In Figure 8, line graphs represent voltage across an inductor over time as the field is rising and as it is falling. The increments in the axes are even (every 20 mA, 2 V, and 10 ms), the origins are zero, and the plotted points are rounded out to make a curve.

Pie charts are partitioned circles in which each partition represents a percentage or proportion of the category. Pie charts should be drawn with a compass for a perfect circle and an exact center. The first segment usually begins at a line from the center to the top of the circle. The segments should move clockwise from the largest to the smallest segment. If there are more than five segments, label the last segment "others" to include all the remaining segments, and itemize the "others" below the circle. Print explanatory information horizontally inside the segment, if possible. If the segment is small, draw a line from it to a space outside the circle and explain it there.

The pie chart in Figure 9 shows the consumption of multilayer printed wire in the United States. According to Gnostic Concepts, Inc., the largest user of this wire is in the computer industry, which uses 52%. The remainder of the wire is used by the communications industry (26%), the government and military services (15%), and other industries (7%). The other industries are not itemized, but they could be. The total sales amount is given, and simple multiplication would break that figure down by section.

FIGURE 8 Line graph of waveforms. From Timothy J. Maloney, *Electric Circuits: Principles and Applications,* © 1984, p. 329. Reprinted with permission of Prentice-Hall, Inc.

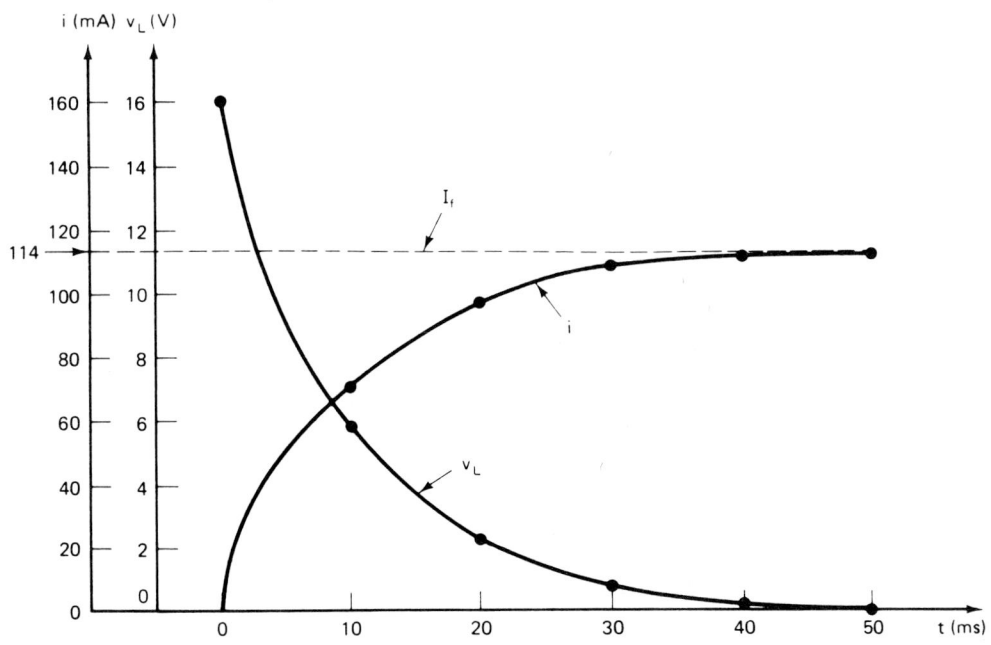

Vertical scale of rising curve: 20 mA/cm

Vertical scale of falling curve: 2 V/cm

Horizontal scale (sweep speed): 5 ms/cm

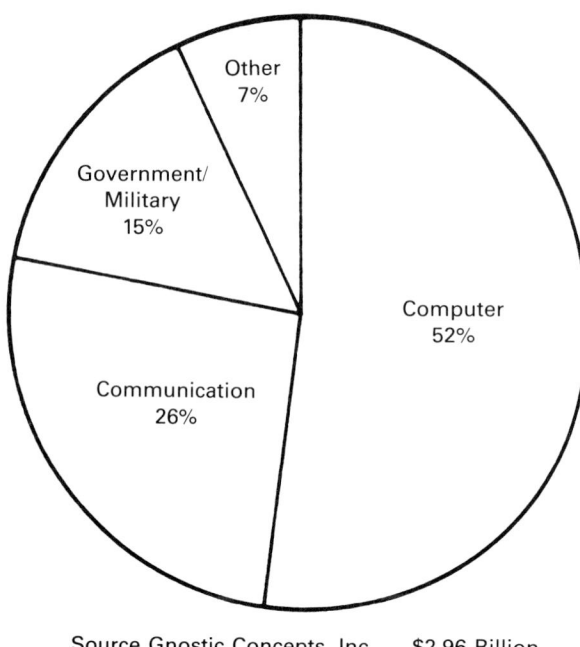

FIGURE 9 Pie chart of U.S. consumption of multilayer printed wiring (1989) by end use. Reprinted with permission from *PC FAB*, June 1985.

TABLES

Tables are displays of information in columns and rows. There is no limit to the number of columns or rows, and the values within the tables are precise. Tables are used to compare and contrast the features of two or more objects. Information is classified and organized into column headings. Columns are arranged horizontally with vertical lines between the columns. A subtotal or grand total can be placed at the bottom of a column and is emphasized with one or two horizontal lines separating it from the rest of the column.

1. Draw a box around the tables to separate them from the rest of the report.
2. Label the table at the top of the box. Include a title that explains the contents of the table. Cite the source at the bottom of the box.
3. Label each column. Provide adequate space between columns to accommodate the longest value in each column. Capitalize all major words in column labels.
4. Keep values listed in a column in a consistent form (if one value includes cents, all values should include cents).
5. Use division lines (horizontal or vertical) to group columns or rows if they make the organization more understandable. Remember, however, that too many divisions make the organization unclear.

Table 1 shows the actual points for the falling exponential curve displayed in Figure 8. The voltage of each of the five time constants is given, including the zero reading at 50 ms.

TABLE 1 Waveform Values
Source: Timothy J. Maloney, *Electric Circuits: Principles and Applications,* © 1984, p. 329. Reprinted with permission of Prentice-Hall, Inc.

Number of Time Constants	Actual Time (ms)	Percent of Final Voltage (%)	Actual Voltage (V)
1	10	37	5.92
2	20	14	2.24
3	30	5	0.80
4	40	2	0.32
5	50	0	0

LET THE READER BEWARE: FIGURES CAN LIE

It is important to be aware that graphical representations of data can be misleading. The graphics can be constructed in ways that visually distort the data without being exactly dishonest. Books, newspapers, and technical journals will occasionally present line and bar graphs, for instance, that appear to offer unmistakable evidence of something, when, upon closer inspection, the evidence is unconvincing or nonexistent. The following examples are only a few of the methods used to misrepresent data.

In Figure 10 the origin of Figure 7 has been changed from zero to $5 million. By not starting at zero, the differences are overemphasized.

In Figure 11 the line graph uses a large increment (amperes rather than milliamperes) to deemphasize the difference.

In Figure 12 the line graph suggests a relationship where the cause and effect are not clear, or that may be linked by other variables, such as population.

Examine all graphics carefully, particularly when you are using them to draw a conclusion. Prepare your own graphics with integrity.

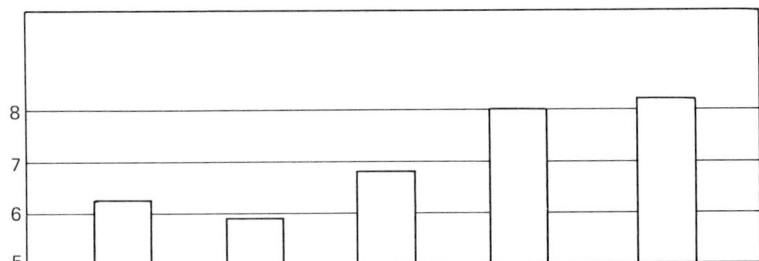

FIGURE 10 Origin of 5 (rather than zero) overemphasizes differences. Compare with Figure 7.

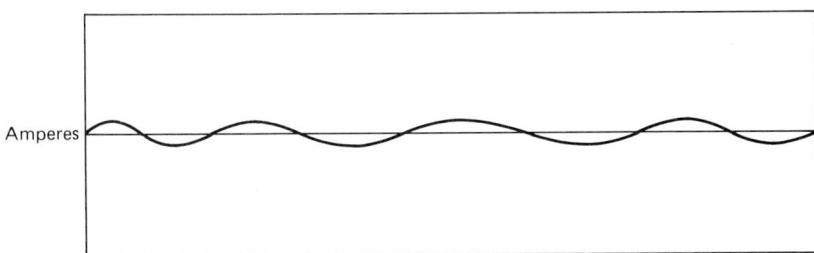

FIGURE 11 Ampere units (rather than conventional milliamperes) deemphasizes differences.

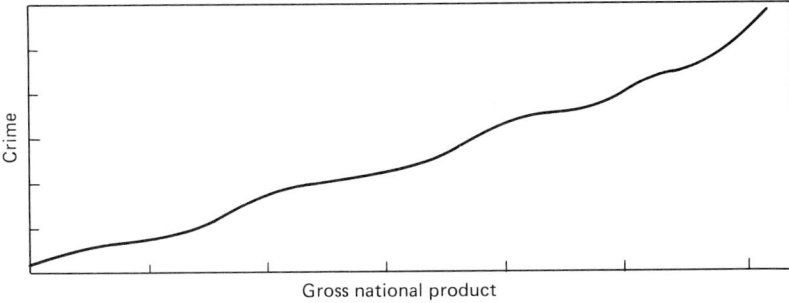

FIGURE 12 Unrelated cause-and-effect relationship.

124 Preparing Graphics

Exercise 10.1 Using what you have just learned about graphics, complete each assignment.

A. Using Figure 13, write a functional description of a multimeter. Refer to the numbers in your description.

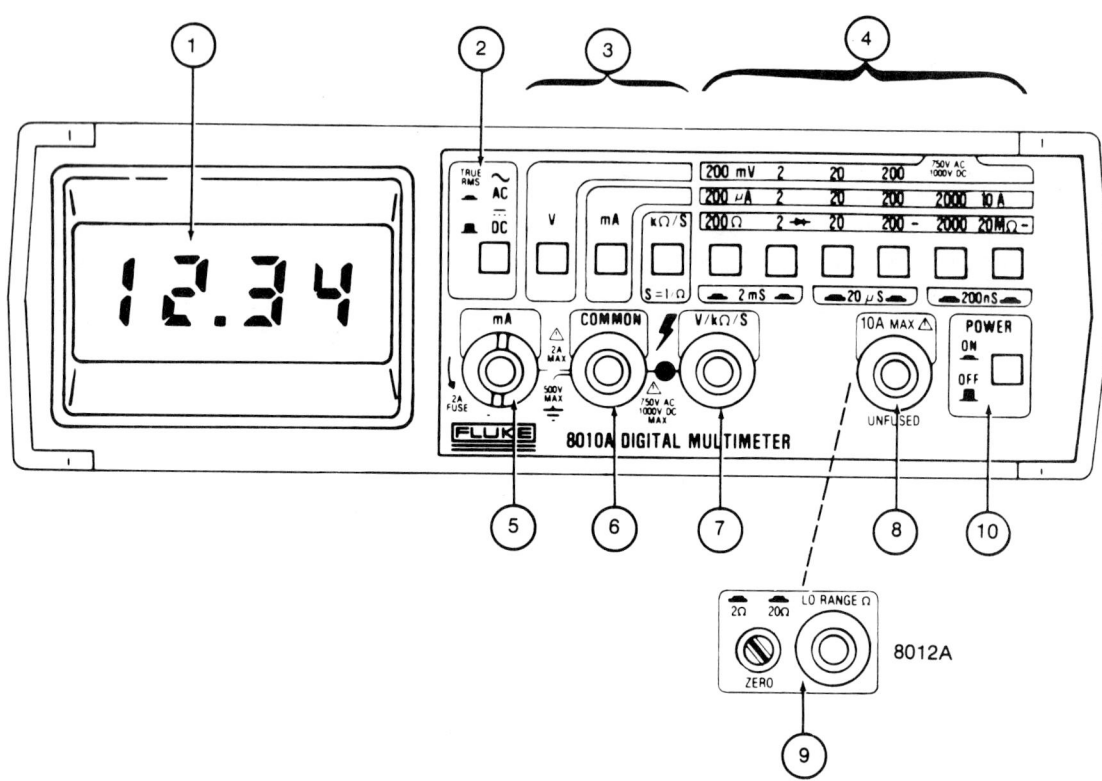

FIGURE 13 Multimeter control panel. Courtesy of John Fluke Manufacturing Company, Inc.

B. Draw, color, and label a resistor to describe how the color code is used to determine its resistance value. Write a descriptive paragraph titled "Using the Color Code" that refers to the drawing.

C. Draw a block diagram of one of the following: a radio, a stereo set, or an IC chip. Remember to include inputs, processes, and outputs. Write a descriptive paragraph explaining the blocks.

D. 1. Complete a table to compare the following information from the Business Communications Company.

   The U.S. printed circuit board (PCB) market in 1985:
   Double-sided—54%, multilayer—32.7%
   Single-sided—8.3%, flexible—5%
   Total sales—$5.41 billion

   The projected U.S. PCB market for 1990:
   Double-sided—52.7%, multilayer—38.3%
   Single-sided—5.3%, flexible—3.7%
   Total sales—$10.4 billion

   What conclusions can you draw from these statistics?
   2. Draw pie charts to represent the statistics in part 1.

E. Draw a bar graph and a line graph to display the following information.

　　　According to the Acme Business Bureau, the Hot Circuit Company
　　has experienced the following annual sales:
　　　　In 1979, its sales totaled $2 million.
　　　　In 1980, its sales totaled $8 million.
　　　　In 1981, its sales totaled $21 million.
　　　　In 1982, its sales totaled $50 million.
　　　　In 1983, its sales totaled $74 million.
　　　　In 1984, its sales totaled $103 million.
　　　　In 1985, its sales totaled $97 million.
　　　　In 1986, its sales totaled $95 million.

　　Write a one-paragraph description of the eight-year sales history of the company.

F. Bring in three examples of graphics from a newspaper or magazine article or advertisement. Write a one-paragraph interpretation of each graphic and include an evaluation of its purpose, effectiveness, and honesty.

SPELLING: Double Trouble

In this unit, rather than review a spelling pattern, we are going to concentrate on certain technical words that present writers with spelling problems because of a troublesome double consonant. Doubled letters are unvoiced, so we have no audible clues to remind us of the letters.

The only way to remember the correct spellings of these words is to practice spelling them correctly and observe them carefully to form a mental picture of the words. Soon you will recognize when a misspelled version "doesn't look right." Whenever you are unsure of a spelling, use the dictionary.

Exercise 10.2　Draw a box around the double letters in each word. Then write the word twice.

1. accomplish _____
2. antenna _____
3. approximate _____
4. assemble _____
5. attenuator _____
6. battery _____
7. collapse _____
8. collector _____
9. communicate _____
10. connect _____
11. current _____
12. dissipated _____
13. efficient _____
14. emitter _____
15. oscilloscope _____
16. parallel _____
17. personnel (people) _____
18. profession _____
19. symmetrical _____
20. transmission _____

Exercise 10.3　Pick out five words from the list above that you had trouble spelling, and use them in a sentence.

1. _____
2. _____
3. _____
4. _____
5. _____

Exercise 10.4 Classify the words above according to their part of speech. Do not change word endings. Some words could fall in two categories. Be prepared to justify your choice.

Nouns	Verbs	Adjectives

Exercise 10.5 From the list of words above, fill in a word to complete each sentence. Each word may be used only once or not at all.

1. The voltage across each branch of a _____ circuit is the same.
2. The amount of power _____ in a resistor is proportional to the square of its _____.
3. A _____ is a dc voltage source consisting of one or more cells.
4. A _____ line that quickly and accurately transfers signal energy from one location to another is said to be _____.
5. The _____ is a device or circuit used to reduce the intensity of a signal without distortion.
6. Wire is used to _____ two points in a circuit.
7. The technician carefully read the directions on how to _____ the power supply.
8. The _____, the base, and the _____ are the three terminals of a bipolar transistor.
9. Exact, rather than _____, readings are displayed on the _____ screen.
10. The decision to hire the graduate was approved by the _____ manager.

VOCABULARY: Tele/Phono/Photo Graph/Gram

The study of electronic communications frequently uses five Greek root words. TELE means "far off" or "at, over, to, or from a distance."

  telecommunication
  telescope

PHONO or PHONE means "sound, tone, or speech."

  telephone
  phonograph

PHOTO means "a light" or "produced by a light."

> photograph
> telephoto lens

GRAPH means "something that writes or records" or "something written."

> graphics
> telegraph

GRAM or GRAMMA means "something written down or recorded."

> grammar
> electrocardiogram
>
> *Note:* Do not confuse GRAM, the root word, with GRAM, the metric unit of weight.

Exercise 10.6 Write a brief meaning of each example word as it relates to the Greek root. Use the dictionary only as a last resort.

Example: telecommunication—communicating from a distance

1. telescope _____
2. telephonic _____
3. phonograph _____
4. photograph _____
5. telephoto lens _____
6. graphics _____
7. telegraph _____
8. grammar _____
9. electrocardiogram _____
10. graphics communication _____

Exercise 10.7 Predict the meanings of the following words as they relate to the Greek roots. Use the dictionary or textbook only as a last resort.

Example: telemetry—measuring from a distance

1. diagram _____
2. teletype _____
3. photoelectric cell _____
4. telemeter _____
5. phototransistor _____
6. radiotelegraph _____
7. phonogram _____
8. graphite _____
9. photoemission _____
10. photosensitive _____

WORD WATCH: Lose/Lost/Loss/Loose/Loosen

The words LOSE, LOST, LOSS, and LOOSE can be confusing to writers who hurry through their first drafts and do not bother to proofread. With attention and a crutch, you can be sure of the correct use of each of these words.

LOSE (pronounced *looz*) is a present-tense verb meaning "misplace," "be deprived of," "give up," "waste," or "bring to ruin." The other tenses are LOST (past, past participle) and LOSING (present participle). There is no such word as "losed."

> I tried not to *lose* the phone number.
> I *lost* it anyway.
> I *have lost* several valuable phone numbers.
> I think I *am losing* my mind.

LOST can also be an adjective describing something that is hopeless or missing.

> It seems to be a *lost* cause.
> I had to make up for *lost* time.

LOSS is a noun meaning something that is lost. If *a, an,* or *the* can be placed in front of the word, use the noun form.

> The missing phone number was a terrible *loss*.
> The company took a *loss* on the sale.

LOOSE (pronounced *loos*) is completely unrelated to the other similarly spelled words. It is most commonly used as an adjective meaning "free" or "unrestrained." As a crutch, think of TOO LOOSE (both have OO).

> I searched through my *loose* change.
> We finished up all the *loose* ends.

LOOSEN is the verb form of "loose," meaning "the act of making something free or unrestrained." The tenses are LOOSENED (past, past participle) and LOOSENING (present participle).

> I always *loosen* my tie when I arrive at work.
> I *loosened* my seat belt after the plane took off.
> I *have loosened* all the screws on the casing.
> While I *was loosening* the screws, the case fell off.

Exercise 10.8 Use LOSE, LOST, LOSS, LOOSE, LOOSEN, or LOOSENED to fill in the missing words in the paragraph.

The factory experienced a power _____ every day at 10 a.m. About one hour was _____ every day, and the management was at a _____ to explain the mysterious "downtime." The maintenance staff _____ all the power outlets to look for _____ connections. This resulted in even more _____ time, but the foreman decided that it was more cost-efficient to _____ a few hours while methodically troubleshooting and correcting the problem than to suffer an ongoing _____.

Exercise 10.9 Use each of the following words in a sentence.

1. lose _____
2. lost (verb) _____
3. have lost _____
4. lost (adjective) _____
5. losing _____
6. loss _____
7. loose (adjective) _____
8. loosen _____
9. loosened _____
10. loosening _____

CHAPTER ELEVEN

PROCESS REPORTS

- *Write a set of directions.*
- *Expand the directions into a process report.*
- *Present an oral report of a process.*
- *Spell ise/ize/yze endings correctly.*
- *Use micro/macro prefixes correctly.*
- *Use advice/advise correctly.*

READING: Space Technicians Service the Satellites

For almost as long as there's been electronic equipment to break down, there have also been service technicians to come out and fix the wayward equipment. Now most service technicians probably have their own tales about a *way out* service call, but none will top the one that can be told by Drs. George D. Nelson and James D. van Hoften. After all, their *service call* took them nearly 270 miles, straight up.

The Solar Max

On February 14, 1980, the Solar Maximum observatory (or Solar Max, as it is commonly known) was launched. Its mission was to provide solar scientists with information about the sun that was impossible to obtain using earthbound instruments.

Among the subjects to be studied by using the satellite were solar flares. Radio amateurs are familiar with the effects caused by those massive solar storms. Within hours after the occurrence of a flare, earth-bound communications can be severely curtailed and brilliant auroras (northern lights) may be generated.

One of the things known about solar flares is that they appear to occur in cycles, with peaks in activity falling about every 11 years. One such year of peak or solar maximum activity was 1980; thus, the Solar Max was launched in that year to learn more about the sun in general, and solar flares in particular, during the active period.

Solar flares are complex phenomenon. In fact they are so complex that no single instrument can monitor all of their features. Because of that, Solar Max carried seven scientific instruments that are simultaneously used to monitor a flare. Six of those instruments are used to measure the energy emitted in the electromagnetic

CARL LARON

High above the clouds and atmosphere astronauts serviced Solar Max. It's the first in-space satellite service call!

"Space Technicians Service the Satellites," by Carl Laron. Reprinted by permission from *Hands-On Electronics*, Spring 1985, pages 22–26. Copyright © 1985, Gernsback Publications, Inc.

spectrum, from visible light to ultra-violet, and from X-rays to gamma rays. The data gathered by those instruments are used to help us learn more about how flares start, how they release their energy, and how that energy affects the earth. The seventh instrument on board the satellite is used to measure the total radiation released by the sun.

Trouble in Space

During its first nine months in space, the Solar Max mission was a smashing success. Then, suddenly, three fuses in the satellite's Attitude Control System blew, causing the satellite to lose its ability to point its onboard instruments accurately. That rendered all but three of the instruments useless. Later, a fifth instrument developed an electronic problem and failed.

Because the mission objectives established before the launch were met, the Solar Max project was declared a success. Nevertheless, since the lifetime of the satellite was reduced from an expected two years to just nine months, a great deal of useful data was lost to earthbound scientists.

But Solar Max was one of the first of a new breed of satellites: one that was, at least in part, designed to be repaired in space. The faulty Attitude Control System was one of three replaceable modules. Those modules, which are used for such things as power and positioning, are located in the bottom portion of the satellite (see Fig. 1). They are standardized units, called the multimission modular system, and are designed to be used in a wide variety of satellites.

When the space shuttle program was first conceived, one of its primary benefits was intended to be the repair or retrieval of satellites. But until it was actually accomplished, no one could say for sure

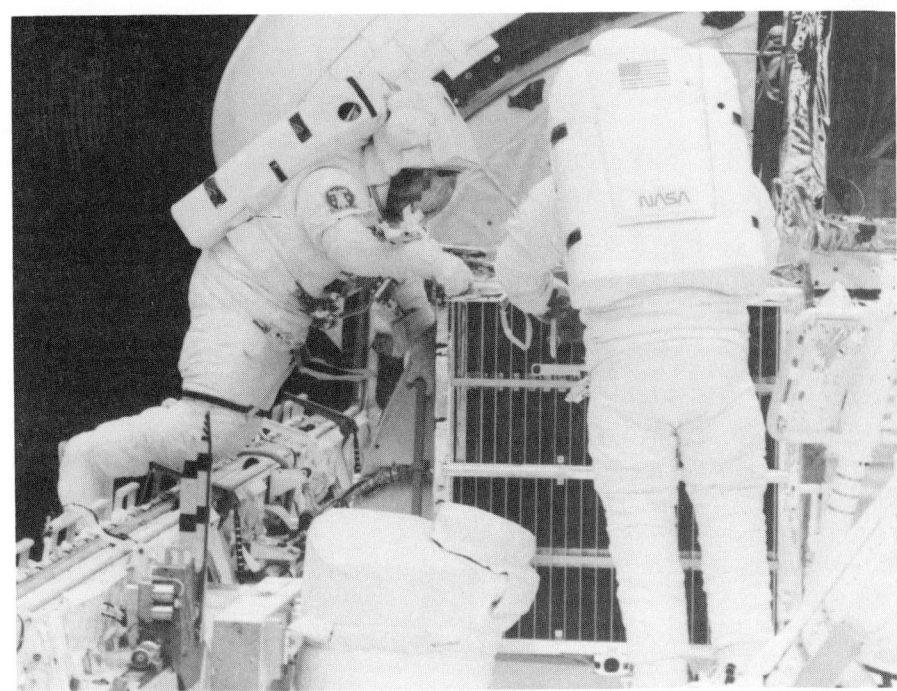

Astronauts James D. van Hoften, right, and George D. Nelson, 41-C mission specialists, work cautiously to change a faulty attitude control module on the Solar Maximum Mission Satellite "captured" in the aft end of the Challenger's cargo bay. Dr. van Hoften is anchored to a "cherry picker" device which involves a foot restraint/workstation connected to the remote manipulator system of the Earth-orbiting spacecraft. Courtesy of the National Aeronautics and Space Administration.

FIGURE 1 An exploded view of the Solar Max satellite. The three modular units at the base of the satellite are intended for replacement while in space. One of those units, the Attitude Control System (ACS) was replaced in space by Drs. James D. van Hoften and George D. Nelson.

1. Coarse Sun Sensors
2. Thermal Enclosure
3. Electronic Enclosure
4. Instruments
5. Instrument Support Plate
6. Solar Array System
7. Attitude Control System (ACS) Module
8. Transition Adaptor
9. Module Support Structure
10. Trunnion Pin
11. Grapple Point
12. Power Module
13. C and DH Modules
14. High-Gain Antenna System (HGAS)

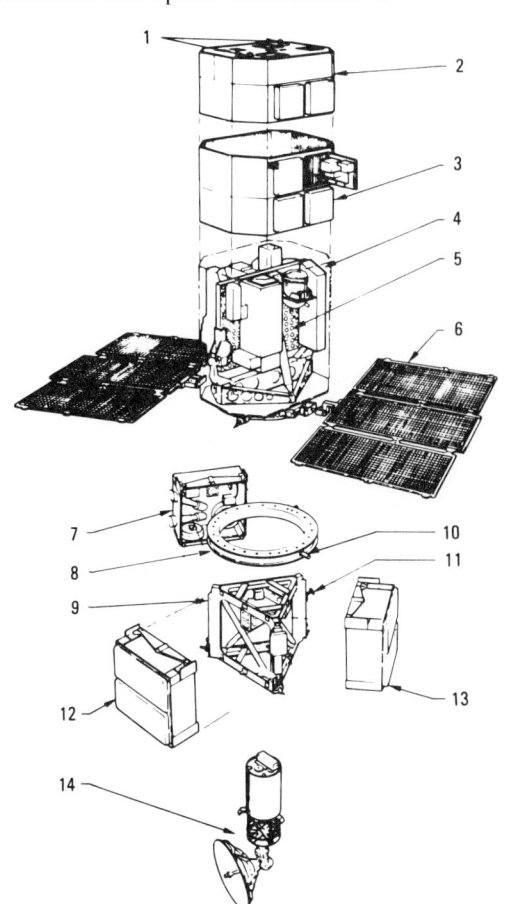

that such a feat was possible. Because of its design, and relative importance to earthbound solar scientists, Solar Max was chosen to be the first test of that capability.

A Service Call

The space shuttle *Challenger*, with a crew of five, was launched from the Kennedy Space Center in Florida on April 6, 1984. Among its prime mission objectives was the repair of the Solar Max. That repair was to consist of two parts.

One part was the replacement of the Attitude Control System. The second, and more difficult repair was the replacement of the main electronics "box" in the Coronagraph/Polarimeter, an instrument that was used to study the sun's outer atmosphere, or corona, by creating an artificial eclipse. That instrument was not intended for repair in space, and the faulty circuitry was located inside the insulated shell of the satellite; but the instrument was positioned in such a way that it was felt that a repair was feasible.

For any repair job the proper tools are needed, and the repair of a satellite in space is no exception. One of the most important tools used in the repair is the Flight Support System, a mounting platform for the satellite during repairs (see Fig. 2). The support system was designed to mate with the multimission modular base of the satellite.

The Flight Support System is U-shaped and fills the 15-foot width of the shuttle's cargo bay. It consists of three cradles, called A, B, and A prime, and a circular ring to which the satellite can be anchored during the repair.

The Flight Support System provides electrical and mechanical connections between the shuttle and the satellite under repair. A rotator on the support system can turn the satellite, and a pivotor is

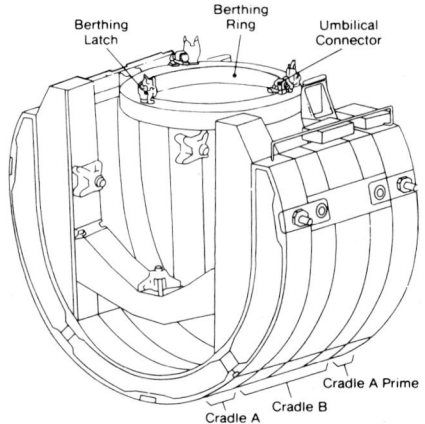

FIGURE 2 During repairs, the satellite is mounted in the Flight Support System cradle. That U-shaped cradle fills the width of the shuttle's 15-foot cargo bay.

used to tilt the satellite at any angle from upright to horizontal. To secure the satellite, three berthing latches clamp onto pins near its bottom. Power and heat from the shuttle are provided to the satellite via umbilical connectors.

A module service tool is used to remove the modular Attitude Control System. That module service tool, which has been specially designed for that task, is controlled by two handles and two switches. To remove the faulty module, two latches on the tool are inserted in matching holes in the module. The latches are used to hold the tool in place, as the socket wrench at the end of the device is used to loosen the two retention bolts that hold the module to the satellite. (If a regular socket wrench were used on those bolts, the astronaut would turn—but the bolts would not move!)

Once the bolts are removed, the service tool is used to hold the module and it is slipped off the satellite. The faulty module is then stowed in a temporary location; the new module is retrieved and mounted using the tool; and the faulty module moved to its permanent location for return to earth. On earth, the module will be examined to find the cause of the failure and may possibly be repaired for future use.

One of the problems of repair in space is that while objects are weightless, they are not massless. The modules are large units, measuring some 4 × 4 × 1.7 feet. On

Solar Maximum Mission Satellite orbits through space, providing no useful function due to the on-board malfunction. Courtesy of the National Aeronautics and Space Administration.

earth they would weight 500 pounds. Because of that, the astronauts doing the repair must work carefully and slowly. If one of the modules were set in motion by accident, it would be hard to control and could do considerable damage.

The Coronagraph/Polarimeter unit is not a module, and its repair is more difficult and time-consuming. But the "tool kit" used in its repair might be a bit more familiar to earthbound technicians. It consisted of scissors, adhesive tape, and an electric screwdriver. To do the repair, the astronaut must first cut through the foil insulation and remove the screws that secure a protective thermal blanket over the instrument. After taping the insulation and blanket out of the way, a panel is unscrewed and opened, exposing the main electronics "box," which is about the size of a briefcase. To complete the repair, the faulty box is disconnected and removed; the new one installed and reconnected; the panel closed, and the protective insulation reattached.

Best Laid Plans

As anyone who has ever serviced electronic equipment knows, rarely do things go as planned. And that service call was no exception.

When the satellite failed, ground controllers placed it in a slow, stable spin of about one degree per second. That kept the satellite's solar panels constantly pointed toward the sun, keeping the unit's batteries fully charged.

As you might imagine, grabbing a spinning satellite is no simple matter. That's because, even though it is weightless, the satellite retains its momentum.

The first task, then, was to stop that spin so that the shuttle's manipulator arm could pick up the satellite safely. To do that, an astronaut—in this case Nelson—was to fly over to the satellite using a jet-powered Manned Maneuvering Unit, or MMU. Once at the satellite, Nelson was to stop the spin by docking himself to a pin protruding from the Solar Max and using the maneuvering unit's 1.7 pound thrusters. To accomplish the docking, a special receptacle, called the Trunion Pin Acquisition Device, TPAD, was mounted on the arms of the MMU.

By the mission's third day, the *Challenger* had caught up with the wayward satellite, and astronaut Nelson set out for his short 10-minute trip to Solar Max on schedule. When he reached the satellite, he matched its spin by firing the MMU thrusters. So far, so good.

Then trouble hit. When he attempted the docking, the jaws within the TPAD failed to clamp onto the satellite's pin as they were supposed to. Nelson tried again, this time moving in a bit faster, but still could not dock.

The failed docking attempts caused two problems. The first was the uncontrolled spin; the second was that the solar panels were no longer aimed toward the sun. Because of the latter, the satellite's batteries would power the satellite for only an additional six hours.

Fortunately, using telemetry, ground controllers were able to stabilize the spin. In fact, they were able to stabilize it to the point where the satellite was holding almost perfectly still. And, more fortunately still, though the panels were still pointed away from the sun, the satellite's orbit swung around and pointed those panels toward the sun just before time ran out on Solar Max's batteries.

By the next morning, the satellite's batteries had almost completely recharged, and it was time to try to retrieve the satellite once more. Because the satellite was turning much more slowly (at about half of its original rate), and because fuel levels aboard the shuttle were beginning to run low, it was decided to try using the manipulator arm once again.

By now, the docking attempts had begun to disrupt the satellite's orderly spin. As a final attempt, Nelson tried to steady the satellite by grabbing onto one of its solar panels. That, however, only made matters worse. Since the MMU's fuel supply was now running low, Nelson returned to the shuttle.

But damage had been done. Now, instead of spinning in an orderly manner, the satellite was spinning unpredictably about all three axis. After Nelson returned to the *Challenger*, four attempts were made to grab the satellite's pin using the shuttle's manipulator arm. All of those attempts also failed.

After donning spacesuits, Nelson and van Hoften made their way out of the ship and into the cargo bay, where they immediately went to work.

Their main task, replacing the faulty Attitude Control Module, took about 45 minutes and went smoothly. Replacing the Coronagraph/Polarimeter electronics was supposed to be a lengthy task. After all, the job entailed pulling back a panel covering the box, cutting and taping back a layer of insulation, removing some two dozen screws, and cutting a number of wires, all while wearing bulky spacesuit gloves. All went much smoother than expected, however, and Nelson completed the job, using clips instead of screws to reconnect the severed wires, in less than an hour; pre-flight estimates had pegged the repair time at three hours. The final step in the repair of the satellite was the installation of a baffle cover over an instrument called an X-ray Polychromator, used to measure the X-ray emission from solar flares, to vent its exhaust gas away from the rest of the satellite's instruments.

A Nice Catch

The fourth day of the mission was used to attend to other activities not related to the repair of Solar Max. But, early on the fifth day, the shuttle successfully grabbed the satellite and hauled it safely into the cargo bay.

To perform the maneuver, mission commander Robert L. Crippen carefully positioned the shuttle under the satellite so that astronaut Terry J. Hart could extend the manipulator arm upward and grab onto the satellite's pin. This time, they caught it on the first try.

The satellite was then brought into the cargo bay and secured to the Flight System Support cradle. Power was supplied to the satellite through the umbilical connectors, and the satellite was pivoted around so that Drs. van Hoften and Nelson could get at the electronics more easily.

The fifth day of the mission was the day that repairs were performed. Though originally scheduled for two days, because of the problems the crew ran into while catching the satellite, all repairs now needed to be completed in one day.

The Future

The success of the Solar Max repair mission signaled the dawn of a new era in the repair of electronic instruments. Now, failed satellite-based systems can be successfully repaired, and satellites can be realistically returned to service.

Perhaps key to the future of space repair will be the concept of modular construction. New satellites will make increasing use of modular construction, such as that used in the multimission modular spacecraft base. That will make it possible to replace faulty satellite circuitry with relative ease. The failed circuitry can then be returned to earth, and troubleshooting using conventional techniques can be done on the unit. The unit can then be repaired and used on another satellite. For instance, the Attitude Control Module retrieved in the Solar Max mission can be repaired and used again.

When modular construction techniques, as well as the techniques used to repair satellite modules are perfected, expect more and more modular satellites. Eventually, interchangeable modules may be mass-produced and carried as part of an on-going inventory. When a circuit fails, the service technician of the future may be able to select an appropriate module out of inventory, swap the modules in space, and then repair the module in a more conventional environment.

The Solar Max mission was but a first, halting step. But most important, it was a successful step, and one that showed that such repair missions are both possible and practical.

Reading Comprehension Questions

1. What was the mission of the Solar Maximum observatory (or Solar Max)?

2. Why is the Solar Max called "one of the first of a new breed of satellites?"

3. What was the purpose of the Flight Support System?

4. What was the function of the module service tool?

5. What problems resulted from the failed docking?

WRITING: The Process Report

Technical reports that explain how something works or how to make something work are called **process reports.** Process is the noun form of "proceed," moving from one step to another.

Directions are the simplest types of process reports. Directions tell how to perform a task in brief, ordered steps beginning with the first step ("plug in the cord") and ending with the final step ("unplug the cord").

A **report** is longer and supplies more information than just the steps to get the job done. A report may also contain why each step is performed, what happens as the step is performed, or a method of evaluating the effectiveness of each step or the completed procedure. The report will contain, in addition to the steps to complete the task, an introduction and conclusion.

The most important rule for writing clear directions and process reports is to understand the process well yourself before beginning to write. This may mean doing research, reenacting the process, or watching the process being performed. Scientists have learned the value of keeping daily accounts or notes while an experiment is being performed. Although notekeeping is time consuming during the experiment or procedure, the notes will make writing the documentation or report faster and more accurate.

The second rule is to organize the notes so that the process can be described simply, logically, and accurately for its intended audience. Whether the paper is written to report results, draw a conclusion, guide someone else, or simply inform people, enough information (including data and illustrations) is included to make the report complete.

Obviously, knowing the audience will make a difference in the amount of information included. Imagine, for instance, that the author of the article "Space Technicians Service the Satellites" wrote an article on the same subject for a group of astronauts or aeronautical engineers. The final article would contain many more details, precise measurements, and jargon that a general audience would find meaningless.

The main sections of a process report are the introduction, step-by-step procedures, and conclusion. Some reports may also contain technical data, graphics, a parts or equipment list, and/or a summary. The article is an example of a process report written to inform a moderately technical audience with an electronics background. The readers are not expected to replicate the process or to analyze its success. The readers are only expected to appreciate the difficulty of a routine task being performed in a highly unusual and unpredictable environment—outer space.

The article opened with background information: a description and explanation of Solar Max, solar flares, and the instruments needed to monitor solar flares. The purpose of the article, to recount the actual repair, is stated next. Then the author provides only those details and procedures of the *Challenger* mission that relate to the Solar Max. The conclusion discusses the implications of this mission for future electronic instrument design and repair. The line drawings were included to illustrate uncommon objects: the Solar Max satellite and the Flight Support System cradle. Many other graphical aids could have been provided, but the author decided these two were essential and the others were not.

Process reports are usually written in the second person (*you*) or third person (*it, they*). Rarely do the authors refer to themselves (*I, we*) in directions or reports. The easiest method is to use the "understood you" as the subject and active, present-tense verbs, as in the following examples.

> Place the diode . . .
> Calculate the values . . .
> Measure the frequencies . . .

Process Reports

Using the third person is also acceptable, but it often means using the passive voice, which necessitates more words and a formal tone (and may be appropriate for longer reports).

> The diode was placed . . .
> The values were calculated . . .
> The frequencies were measured . . .

The following example shows a set of technical instructions for soldering a component.

INSTRUCTIONS FOR SOLDERING A COMPONENT ONTO A PRINTED CIRCUIT BOARD

by Christopher Hill

CAUTION: To prevent personal injury or fire, use caution when handling the soldering iron. Always rest the soldering iron on its stand when it is not in use or when it is cooling after use.

1. Obtain the following items:
    - solder (rosin core only)
    - soldering iron and stand
    - printed circuit board
    - electronic components
    - needle-nose pliers
    - wire cutters
2. Plug in the soldering iron, and rest it on its stand while it is warming up. Allow the iron to warm up for 5 minutes.
3. Insert the component leads through the top of the circuit board. The top side is the side without any foil on it.
4. Mount the component flush with the circuit board by pulling the leads all the way through with the needle-nose pliers.
5. Rest the circuit board on its top so that the leads of the component are pointing upward.
6. With the wire cutters, cut off a 6-inch length of solder.
7. Grasp the soldering iron firmly in one hand while holding the solder in the other.
8. Touch the solder to the component lead.
9. Simultaneously touch the soldering iron to the solder, the foil, and the component lead.
10. Hold the soldering iron in place until the solder melts, usually 5 to 7 seconds.
11. Remove the soldering iron and the length of solder.
12. Allow the lead to cool 10 seconds before proceeding.
13. Repeat steps 7 to 12 for the remaining leads.
14. If the leads protrude more than 1/8 inch from the surface of the circuit board, clip the leads to 1/8 inch, or as much as possible, with the wire cutters.
15. If no more components are to be soldered, unplug the soldering iron and allow it to cool for at least 20 minutes before putting it away.

Reprinted with permission from Christopher Hill, DeVry Institute of Technology, 1986.

Exercise 11.1 Write a complete set of directions for one of the following procedures:

- hooking up a VCR
- installing a tape player in a car
- setting up a campsite
- setting a digital watch
- preparing for a journey

Exercise 11.2 Expand the set of directions from Exercise 11.1 into a process report. Include the following sections:

- an introduction (background, purpose, definitions)
- a parts list (all items needed for the procedure)
- step-by-step instructions (one paragraph per step)
- a conclusion (how to evaluate correct performance)

Exercise 11.3 Write instructions and draw graphics to describe how to pick up and hold a pen in your hand.

Exercise 11.4 Prepare an instructional oral report, using at least one graphic or object, on one of the following topics:

- playing a video game
- performing a magic trick
- assembling a model/object from a kit
- playing a musical instrument

SPELLING: Getting wISE to IZE/YZE

The three final syllables ISE, IZE, and YZE all sound exactly alike—we hear it as "eye-z." We add one of these endings to an adjective and the new word is a noun. The new ending means "to make," so the word "standarize" means "to make standard."

To determine the correct spelling of the ending, there are a few commonsense rules. Use your visual memory for frequently used words, as it will help more than remembering the rules. Always check the dictionary when the word "just doesn't look right." Notice these endings when you read.

Rule 1 Use a final YZE for only a few technical words:

        analyze   paralyze   electrolyze

Rule 2 Use a final ISE when it is part of a word, such as WISE, VISE, RISE, and GUISE.

| likewise | advise | sunrise | guise |
| otherwise | supervise | arise | disguise |

Use a final ISE for words ending in -MISE, -PRISE and some words ending in -CISE.

| surmise | surprise | exercise |
| compromise | comprise | incise |

Rule 3 Use a final IZE for nearly all other words.

  apologize   minimize   organize
  alphabetize   mechanize   symbolize
  emphasize   memorize   visualize

In most cases, when you add suffixes to the verb form of one of these spellings, just drop the final E, keep the YZ, IS, or IZ, and add the suffix.

ANALYZE	ADVISE	ORGANIZE
*analysis	adviser	organizer
analyzing	advisement	organization
analyst	advisory	organizational
*analytical	advisable	organizing

*These words vary slightly from the rule.

——————— WARNING ———————

Some writers attempt to sound formal or technical by adding -IZE to words that are not normally verbs. Although these words often indicate an action, try not to overdo them. Overusing IZE can make the writer sound like a robot—artificial and distant. Notice the difference in tone in the following two sentences.

Mechanical: In an effort to minimize errors, we prioritized and analyzed our objectives and organized our procedures.

Human: To reduce errors, we first established our objectives and procedures.

The same overkill can happen when writers add -WISE to a noun and use the new word as an adverb, as in "timewise" and "costwise."

Mechanical: The procedure was more efficient timewise and costwise.

Human: The procedure's efficiency saved time and money.

Writing that sounds human is easier to read than mechanical-sounding writing.

Exercise 11.5 Using YZE, ISE, and IZE, add the correct ending to each word.

1. Turn the dial clockw_____.
2. Use the oscilloscope to visual_____ the waveforms.
3. To anal_____ the results, correct measurements have to be taken.
4. It helped to priorit_____ the needs.
5. The manager took the situation under adv_____ment.
6. The cable was placed lengthw_____ on the bench.
7. She was thrilled when she was asked to superv_____ the project.
8. If the need ar_____s, more staff will be added.
9. No one was more supr_____d than he was to hear of his promotion.
10. In a schematic drawing, sawtooth marks symbol_____ fixed resistors.

VOCABULARY: Micro/Macro

MACRO is a Greek prefix meaning "long" or "large."

Macroscopic means something that is visible to the naked eye.

MICRO is a Greek prefix meaning "small" or "little" (or a metric unit).

Microscopic means something that is too small to see without using a microscope.

When these prefixes are combined with words, they are usually joined without a hyphen.

Exercise 11.6 Use MACRO or MICRO to complete the words in the following sentences.

1. A _____ circuit is a miniaturized integrated circuit found in a computer.
2. A _____ phone is an electroacoustic transducer that converts sound waves into electric signals.
3. A large, complex entity, such as the universe, is sometimes referred to as a _____ cosm.
4. A miniature universe, such as a pond, is sometimes referred to as a _____ cosm.
5. The branch of economics dealing with the behavior of individual consumers is called _____ economics.
6. _____ economics is the branch dealing with the interrelationships of large sectors of the economy, such as employment and income.
7. The larger of two nuclei is called the _____ nucleus.
8. _____ film is film that is reduced in size after being photographed.
9. A _____ processor is the controlling unit of a microcomputer laid out on a silicon chip.
10. A _____ wave is part of the electromagnetic spectrum of wavelengths ranging from 0.3 to 30 cm.

WORD WATCH: Advice/Advise

In the spelling exercise in this chapter, you saw a crutch to help remember how to use these easily confused words. Remembering how the pronunciation of the word affects its spelling will help you in difficult situations.

ADVISE ends with the sound of "eye-z," and it is a verb that means to give an opinion or counsel. *Advise* is a regular verb (advised, advising), and the other forms are *advisory, advisement,* and *adviser.*

Your *adviser* will *advise* you to join the *advisory* council.

ADVICE ends with, and sounds like, the small word "ice." *Advice* is a noun, meaning the opinion or counsel that is given. It has no other forms. Although a plural form of *advice* is occasionally written as *advices,* it is usually considered an "uncountable" noun, meaning that we do not make it singular or plural.

>I will use the *advice* to guide my career.

Exercise 11.7 Fill in the blank using the correct form, ADVISE or ADVICE. Drop the final E if necessary when adding an ending.

1. Because I don't know anything about the subject, I can't offer you _____.
2. Workers are _____d to wear safety glasses when entering the room.
3. I am too involved in the situation to have any objective _____ about resolving it.
4. The _____ory group was formed to handle the situation.
5. Personnel problems require special _____ment from qualified counselors.
6. He ignored the _____ of his staff and followed his natural instincts.
7. He learned to withhold _____ until it was asked for.
8. Sometimes just talking about a problem with a trained _____r solves the problem.
9. She spent so much time _____ing others that she didn't notice her own problems.
10. _____ is often easier to give than to follow.

CHAPTER TWELVE
REVIEW

READING: Message by Light Wave

ROBERT W. LUCKY

In 100 years we have grown accustomed to the concept of radio transmitted communications, but the concept must once have seemed terribly abstract and mysterious. Now the world of communications has turned its attention to the much more intuitive optical technology, which brings to mind images of smoke signals rising above the forests or flashes of semaphore lamps glimmering across the seas. Daily, we use our own "optical receptors" for communications.

If the capacities of optical fibers were fully exploited, the entire present telephone voice traffic in the United States could be carried on a single fiber. The contents of the Library of Congress could be transmitted in a few seconds. We have apparently discovered an inexhaustible medium for communication, yet fibers thus far have been used largely as direct replacements for electrical equivalents. Our everyday world has been changed little, if at all.

Lightwave communications began in earnest in 1970 when researchers at Corning Glass fabricated a useful lightwave conduit—a glass fiber about the diameter of a human hair. This achievement caught the attention of the communications industry because optical fiber offers an enormous "bandwidth" advantage over electrical transmission.

The bandwidth of a communication channel is a measure of range of frequencies supported—like the spread between the highest and lowest note on a piano. This critical number determines the information-carrying capacity of the medium. The frequency of light is so much higher than that of the traditional radio frequencies that

In fiber optics, a burst of light is worth a million words.

"Message by Light Wave," by Robert W. Lucky, *Science 85*, November 1985, pp. 112–113. Reprinted by permission.

141

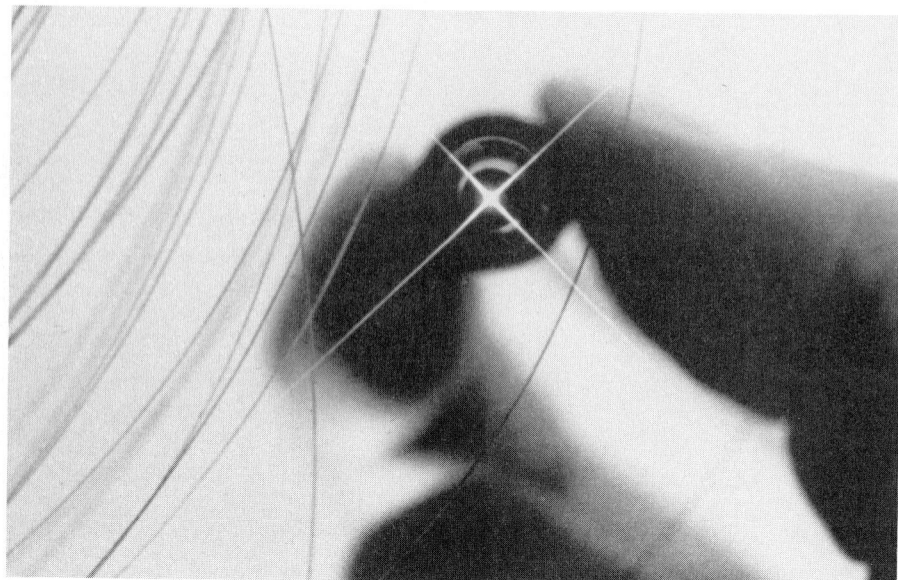

Artistic view of optical fibers and light pulse from end view of fiber optic cable. © 1987 by AT&T; all rights reserved.

the bandwidth of a lightwave fiber is about a million times greater than a radio channel.

The past 10 years have seen dramatic advances in how much an optical fiber can carry and how fast. The transparency of silica glass fiber has approached the theoretical maximum; light pulses can be seen after more than 100-kilometer transmission. The current experimental record is four billion bits—about the information contained in a 30-volume Encyclopaedia Britannica—transmitted each second over a span of 117 kilometers. With the information-capacity limit still perhaps five orders of magnitude away, it is likely this progress will continue through the coming decade.

An entire optical transmission technology, however, has not been developed in a mere decade. The superhighway of optical fiber alone does not make a useful communications system. There must be means to gather traffic, to route diverse streams of information, and to logically package and unpackage enormously fast streams of data. This optical switching and logic is yet to come.

The next step in underlying optical technology should be to develop a collection of miniature light-wave "plumbing" modules—amplifiers, switches, couplers, filters, and isolators—and then integrate these devices into microcircuit chips. The optical plumbing should enable the mixing of many wavelengths of light onto a single fiber. Radio communications has handled this nicely for many years. The AM radio band, for example, contains many different stations, each on its own frequency. Commercial optical fiber systems now generally transmit data on a single wavelength, like a single radio station. The enormous bandwidth of the optical fiber remains unexploited because of difficulty in establishing and manipulating subchannels.

Researchers are struggling with the plumbing problems because the full development of optical technology can come none too soon. The current flow of information is growing too large for electronic processing. Our useful electron will be replaced by the upstart photon.

The first commercial fiber systems linked only telephone offices within metropolitan areas. Then the installation of long-haul fiber systems began with corridors in the Northeast and in California. Before the end of this decade the continental United States will be spanned with high-capacity, competing trunks, super highways of optical fibers.

In 1988 the first fiber submarine cable, connecting the United States, England, and France, is scheduled to go into operation. A similar system is planned for the Pacific, and undersea fibers will link islands with nearby mainlands.

Extension of optical fiber into the home is the next step, one

Spools of optical fiber. © 1987 by AT&T; all rights reserved.

made difficult by economics rather than technology. So far we have built superhighways without on and off ramps. The fiber thruway ends at a distribution point perhaps a mile from home. All that potential bandwidth yearns for the home, but copper telephone wires and coaxial CATV cables serve the market with an economy that fiber cannot presently match. The cost of fiber systems should drop, but even if they do not, fiber could at least multiply the services that wires and cables currently provide. We could have hundreds of television channels. Two-way videotelephone finally could be economical and ubiquitous, and the availability of image material would be so instantaneous that we could browse through television the way we leaf through a magazine.

Currently it is as if the fiber were a superhighway through undeveloped land. Our fiber highway will, like its travel counterpart, undoubtedly draw new businesses, commuters, and even Sunday drivers. Humans have never failed to use the full capacity of available communication technologies. Past its developmental challenges, we will inevitably find new ways of using—and squandering—the potential of optical communication.

Reading Comprehension Questions

1. What two types of communication are being compared in this article?

2. Draw up a comparison and contrast table to highlight the similarities and differences of the two types.

3. Why does the author think optical fibers will continue to advance in the coming decade?

4. What are some current uses of fiber optics?

5. What are some of the problems that need to be resolved before fiber optics can extend inside our homes?

6. Explain the author's analogy of the fiber and a superhighway.

REVIEW: Chapters 9 to 11

Exercise 12.1   Combine the following sentences to make a unified essay. Lable the major sections of the essay:
- introduction
- body headings
- conclusion

Use combining techniques of coordinators, signal words, and punctuation to reduce the total number of sentences. Vary the sentence structure using simple, compound, and complex sentences in each paragraph.

COMMUNICATING BY LIGHT WAVES

Paragraph 1:
We have grown accustomed to the concept of radio-transmitted communications.
The concept must once have seemed mysterious.
Now we use sophisticated technology daily.
We have turned our attention to fiber optics.
We have apparently discovered an inexhaustible medium for communication.
Fibers have hardly been developed.

Paragraph 2:
Fibers, thus far, have been used mostly as replacements for electrical equivalents.
Our world has changed very little.
Fibers are not being used to their full capacity.
The entire present telephone traffic could be carried on one fiber.
An entire encyclopedia could be transmitted in seconds.

Paragraph 3:
Light-wave development began in earnest in 1970.
It caught the attention of the communications industry.
The bandwidth of a communication channel is a measure of the frequency range.
It is similar to the spread between the highest and lowest notes on a piano.
The bandwidth is a critical number.
It determines the information-carrying capacity of the medium.
The frequency of light is higher than that of the traditional radio frequencies.
The bandwidth of a light-wave fiber is greater than a radio channel.

Paragraph 4:
Copper cables and fiber optic cables both transmit messages.
Fiber optic cables transmit information farther and are easier to install.
Voice, image, and data are transmitted by light signals and carried over hair-thin glass fibers.
The progress will continue through the coming decade.
An entire technology cannot be developed in a mere decade.

Paragraph 5:
The production of fibers requires the most exacting conditions.
The production of fibers requires sophisticated machinery.
An optical fiber begins with a 4-foot hollow tube.
The tube is made of pure quartz glass.
A gas is injected into the tube.

The gas sticks to the sides of the core.
The gas will provide the channel.
The channel guides the beams of light through the interior of the strand.
The tube is then heated until it collapses into a solid glass rod.
The rod is about 1 inch in diameter.

Paragraph 6:
The rods are installed on a fiber-drawing machine.
One end is slowly fed into a 4000-degree furnace.
The rod melts.
A tiny strand of glass drops down from the bottom of the furnace.
The diameter of the rod is regulated by a computer.
The computer uses laser beams to maintain the proper thickness.
Monitors detect problems.
The monitors will sound an alarm.
Finally, the strand is wound onto a spool.

Paragraph 7:
Technicians perform extensive tests on the hairlike strands.
The tests detect impurities and aberrations.
All of its test results are stored with each spool.

Paragraph 8:
The strands are bundled into tubes or ribbons.
In one packaging arrangement, up to 12 ribbons are stacked in a plastic tube.
The tube keeps them loosely stacked and free of stress.
The tubes are then supported with stainless steel support strands.
The strands are spirally wrapped around the tube.
An inner plastic sheath is applied.

Paragraph 9:
Next, more steel strands are added for extra strength.
Finally, an outer black, watertight plastic jacket is applied.
The finished cable measures approximately ½ inch in diameter.

Paragraph 10:
The superhighway of optical fiber alone does not make a useful communications system.
Optical switching and logic need to be developed.
The next step will be developing a collection of miniature "plumbing" modules.
The devices will be integrated.
A complete line of light-guide apparatus products includes connector cables and other devices for splicing, terminating, interconnecting, and field testing fiber optic cables.

Paragraph 11:

Long-distance systems and local area networks are common applications.

High-strength optical fibers are being developed.

Eventually, fibers can be used to create transoceanic light-wave cable systems.

The systems will span the Atlantic and Pacific oceans.

Paragraph 12:

Some fiber optics maufacturing companies have expanded their facilities.

They anticipated an increase in demand.

The actual market experiences periodic slowdowns.

Future developments of optical fibers may reach into the home.

For now, telecommunications industries account for most of the fiber business.

Exercise 12.2 Insert the following direct quotes in your essay at an appropriate point. Punctuate the quote correctly. Use a parenthetical note to cite the source.

Robert W. Lucky, "Message by Lightwave," *Science 85*, November 1985, p. 112.

"The current flow of information is growing too large for electronic processing. Our useful electron will be replaced by the upstart photon."

Robert Snowden Jones, "Making Tiny Fiber Optics Cable a Gargantuan Task," *The Atlanta Constitution*, 10/27/86.

"The telephone industry is now the primary market for fiber optics. Long-distance companies, like AT&T, are currently installing thousands of miles of cable. And local telephone companies are beginning to install fiber optic cables in their primary trunk lines."

Exercise 12.3 Include the following figure and table in your essay. Label and write a brief description of each. Credit the source. Include a reference to each within the essay.

TABLE 1 (from "Fiber Optics & Communications Newsletter," 9/86)

	Miles Now in Service	Total Miles Planned
U.S. Sprint	6,200	23,000
AT&T	5,300	10,200
Nat'l Telecommunications Network	4,123	11,160
MCI	2,500	7,000
Regional Networks	2,480	9,126
Total	20,603	60,486

Check your essay with your instructor. Essays will vary.

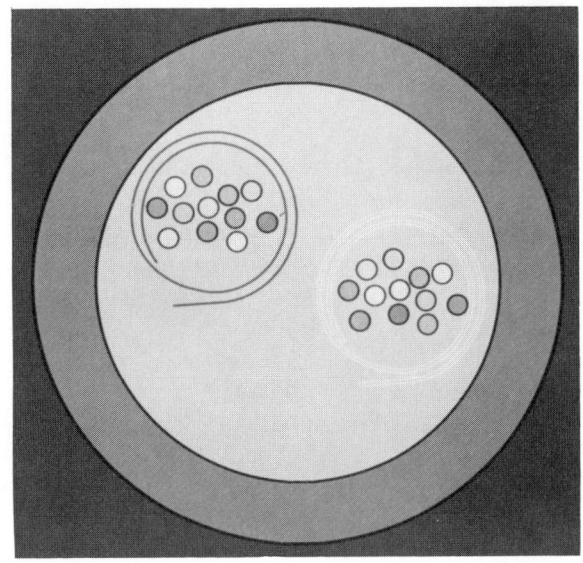

FIGURE 1 A cross-sectional view of a 24-fiber Lightpack cable core tube. © 1987 by AT&T; all rights reserved.

SPELLING REVIEW
Review the following clues. Study the list of words to prepare for a spelling test.

Chapter 9: ANCE or ENCE (Similarly, ANT/ENT) Endings

Clue 1. Use ENCE/ENT when the verb ends in the letter R and the accent is on the last syllable.

      pre-fer' —preference
      re-fer' —reference

Clue 2. Most electronics-related words end in ANCE/ANT.

      resist—resistance, resistant
      resound—resonance, resonant

Chapter 10: Troublesome double letters

Clue 1. Form a visual image of the correct spelling.
Clue 2. Practice spelling frequency used words.

      antenna   oscilloscope   battery

Chapter 11: ISE, IZE, YZE Endings

Clue 1. Use -YZE for only a few technical words.

      paralyze   electrolyze   analyze

Clue 2. Use -ISE when it is part of a final syllable that is a word, such as WISE, VISE, RISE, and GUISE.

      likewise   advise   sunrise   disguise

Clue 3. Use ISE for most words ending in -MISE and -PRISE and most words ending in -CISE.

      compromise   surprise   exercise

Clue 4. Use IZE for most other words.

        emphasize   memorize   organize

Spelling List Practice writing each word.

accomplish	efficiency
advise	inductance
analyze	parallel
antenna	recurrence
attendance	resistance
balance	symbolize
battery	supervise
capacitance	transference
collapse	transmission
dissonance	visualize

Exercise 12.4 Your instructor will give you a spelling test.

Exercise 12.5 Complete the following crossword puzzle using words from the Vocabulary and Word Watch exercises in Chapters 9, 10, and 11.

ACROSS

  6. larger of two nuclei (11)
  9. amateur or beginner (9)
10. electroacoustic transducer (11)
11. positive part of an atom (9)
13. film reduced in size (11)
15. transfer resistor (9)
16. to misplace or give up (10)
20. instrument to see far off (10)
21. prefix: first, original (9)
22. something lost or missing (10)

DOWN

  1. a counsel or opinion (11)
  2. to give counsel or opinion (11)
  3. allows clear light to shine through (9)
  4. the first or original model (9)
  5. something written from far off (10)
  7. a colorless gas (9)
  8. a record of sound (10)
10. prefix: large (11)
11. a record of an image produced by light (10)
12. prefix: across, over (9)
14. prefix: small, little (11)
17. things written or drawn (10)
18. hopeless or missing (10)
19. free or unrestrained (10)

Choose from the following words:

advice	macro	phonograph	telegram
advise	macronucleus	photograph	telescope
graphics	micro	proto	trans
loose	microfilm	proton	transistor
lose	microphone	prototype	transparent
loss	neon		
lost	neophyte		

PART IV

BUSINESS COMMUNICATIONS

Chapter 13 *In-house Communications*
Chapter 14 *Business Letters*
Chapter 15 *Review*

CHAPTER THIRTEEN

IN-HOUSE COMMUNICATIONS

- *Fill out a trouble log.*
- *Write a memo to a supervisor.*
- *Fill out a vehicle report.*
- *Fill out a purchase requisition.*
- *Fill out an expense voucher.*
- *Write an incident report.*
- *Spell and use* ceed/cede/sede *roots correctly.*
- *Use* past/passed *correctly.*

READING: Driving Out the Devils of Communication

PAUL R. TIMM

Pride, covetousness, lust, anger, gluttony, envy, sloth. Theologians say that these seven sins will cause certain spiritual death. And a manager will cause the death of his or her effectiveness by committing the seven deadly "sins" of management communication.

These sins often begin with improper or unhealthy ways of thinking which ultimately lead to wrong behaviors, and we're all guilty of them. But awareness of sin is a great first step to repentance and better management. Let's look at the seven ugly misdeeds that plague us all.

Deadly Sin 1: Not Realizing That Your Message May Get a Different Response Than You Expect. Communication is not a simple process. Very often, other people will react "wrong" to even our simplest messages. This is because people are unique. We each see the world differently based on our individual experiences, values, attitudes, and perceptions. Each of these factors can lead to unexpected reactions when we communicate with others.

In other cases, we get unexpected responses because our message is incomplete. We expect other people to "fill in the blanks"—to read our minds—although needed background information is not supplied or explained. Vague expressions like "Shape up your work area," or "Demonstrate more team spirit" can have widely different meanings to different people.

How can you cope with this first deadly sin? The best advice may be to actually *expect to be misunderstood* and adjust your message anticipating ways your ideas could be misinterpreted.

One simple technique used by professional communicators is

Ways to repent of the seven deadly sins of management communication

"Driving Out the Devils of Communication," by Paul R. Timm, Ph.D., *Management World,* July 1984, pp. 27, 28, 29. Reprinted by permission of *Management World,* AMS, Willow Grove, PA 19090.

repetition. The public address system announcer in a busy airport repeats pages in anticipation of the possibility that the message's target receiver may have missed it the first time. The air traffic controller repeats crucial instructions to reduce the chance of a missed message. Even the best communicators learn as much as possible about their listeners and tailor remarks to their interests, attitudes, and values.

Know your receiver and avoid the first deadly sin of management communication.

Deadly Sin 2: Impressing Instead of Expressing. All communication is a combination of impressing and expressing. While we all want to sound good, look good, or appear to be intelligent, some people get carried away. They impress (or so they think) at the cost of being clear. They fail to adjust language to their audience, to word ideas in ways that others can readily understand. Two factors lie at the root of this sin—failure to recognize differing purposes of different forms of communication, and a desire to exert power.

Communication takes many forms and fulfills many purposes. For example, literary writing is very different from functional business writing—in fact, comparing literature to business communication would be like comparing culture to agriculture. Indeed, in business, we are far more concerned with the yield, and less concerned with sounding impressive, creating elaborate wordplays, and conveying aesthetic qualities. We should have a very clear picture of what we want our reader or listener to do or think.

Another root cause of this sin is that our desire to exert power overwhelms our desire to express ourselves clearly. We're more concerned with sounding impressive and appearing to be knowledgeable, and less concerned with making sure the idea gets across. The power problem often becomes obvious when the writer or speaker tries to dazzle us with jargon, specialized terms, or acronyms, without defining these.

The irony is that you don't really impress people when you confuse them. The most intelligent, thoughtful and effective managers are those who communicate ideas clearly, in simple language.

What's a coping strategy to avoid this sin? Tailor your language to your audience. And, when in doubt, use simpler language. One way to practice using simple language is to explain ideas to children using words that they'll be sure to understand. Ask for feedback to see how well you're doing—kids are great teachers.

Deadly Sin 3: Choosing the Wrong Medium. As managers, we have a variety of communication media available for our use: telephones, memos, letters, interviews, group meetings, and so forth. Too often, however, we get in the habit of using the same medium over and over again. Some people just love to shoot off memos while others call meetings to discuss every issue, or use the telephone extensively but never write a letter.

The best managers are aware of the advantages and drawbacks of each medium. They know, for example, that a letter of commendation is more formal, and thus more powerful, than a casually spoken "attaboy." They recognize that sensitive matters are better discussed one-on-one and that large-audience oral presentations are not a good medium for a problem-solving discussion.

What can we do to avoid the sin of media misuse? One way is to remember the difference between communication efficiency and effectiveness. Communication efficiency is a simple ratio between the cost of a particular message and the number of people "reached." We increase efficiency by cutting the message's cost (for example, running off multiple copies rather than typing original letters) or reaching more people (broadcasting a message to all workers in an office rather than selected audiences).

Communication effectiveness is something different. A message is effective when it reaches its intended audience; is understood in essentially the same way by receiver as intended by the sender; is remembered for a reasonable time; and is used when appropriate.

The important point is that careless media choices may result in high efficiency and low effectiveness. A mass announcement (high efficiency) dealing with a sensitive issue that could better be dealt with through an individual discussion (low efficiency) can result in low effectiveness. Indeed, the most efficient media are often least effective, and vice versa.

Deadly Sin 4: Failing to Close the Feedback Loop. Communication must be multidirectional to be complete. Talking without getting some feedback can hardly be considered communication at all. Failing to close the feedback look can take several forms: (1) not listening to our own messages, and (2) not listening to our receiver's feedback.

An interesting study recently released by the Washington-based Direct Selling Education Foundation confirms the value of feedback, especially from a business point of view. Conducted by the Technical Assistance Research Program, also based in Washington, the study found that 96 percent of unhappy customers do not complain directly to the business, but instead do more damage by complaining to friends and business associates who may be customers or potential customers.

While many people feel it may not be worth their time or effort to complain, figures show that on the average an unhappy customer will tell nine to 10 other individuals about a bad experience, and 13 percent will tell more than 20 other people.

The study also showed that dissatisfied customers who remain silent are least likely to deal with the company again whereas 95 percent of those who complained and received some form of response would again deal with the business if their complaints were handled well and quickly. Statistics also reveal it costs five times as much to make a new customer as it does to keep an existing one—an expensive way to do business.

The telephone company commercial, "Reach out and touch someone," is good advice to managers if worded slightly differently—to avoid this deadly sin, reach out and *listen* to someone—often.

Deadly Sin 5: Applying a Nonverbal Veto. When an unhappy employee bellows, "I'm not mad!" it's pretty obvious that what he or she is saying and what he or she is communicating are two different things. Indeed, studies have shown that as much as 78 percent of meaning is transmitted nonverbally—that is without words. Much of the meaning we convey to other people, we convey through our tone of voice, appearance, timing, and many other nonverbal factors.

There are four essential functions of nonverbal messages: (1) to accentuate information conveyed verbally (gestures, loudness); (2) to express like or dislike (tone of voice, facial expression); (3) to convey intensity of feeling (voice, eye contact); and (4) to contradict verbal messages (the "nonverbal veto").

The adage that actions speak louder than words is being shown empirically to be true. Indeed, when individuals are faced with the choice of either believing the words spoken, or the nonverbal messages associated with the words, they will inevitably believe the nonverbal. Don't create confusion and dilute your verbal messages with nonverbal cues which veto you.

How can we avoid deadly sin number five? Your coping strategy probably should revolve around development of sensitivity to nonverbal characteristics of communication. Think about such things as tone of voice, timing, and mannerisms which may distort your messages. And above all, remember that what you do speaks louder than what you say.

Deadly Sin 6: Not Helping Your Reader or Listener Get the Message. With most messages we need to convey, there are kernels of information which are important, indeed necessary for the reader to understand, and other verbiage which provides the packaging around these key ideas. To help your reader or listener better get the message, you need to help him or her identify the kernels of information.

Two approaches can help you do this. First, try using a preview to create clear expectations in the mind of your message receiver. For example, the memo that begins "In this memo I will be discussing two changes in vacation benefits" does a good job of setting up the expectations of the reader. Likewise in an oral presentation, if the speaker says, "We will cover the three major changes in our compensation plan," he or she is providing a clear content preview.

Another approach is the use of "access" to point to key ideas. In written communication, accessing means using enumeration, white space, short paragraphs, highlighting, and bullets to identify bits of information. No one wants to wade through a large blob of words on a page. Short paragraphs and other access techniques permit a reader to digest one bite at a time.

To avoid deadly sin number six, get rid of the "Agatha Christie syndrome." Business communication should not be written like a mystery novel, with a surprise twist at the end. You should tell the reader what's coming up and make it easy for that person to find your key ideas by accessing them.

Deadly Sin 7: Viewing Communication As a Fringe Benefit. Too frequently, managers consider effective communication as some sort of fringe benefit for employees, as a way to boost their morale. In reality, communication is the essence of the manager's job.

Management is a process of accomplishing tasks through people under the most economical conditions and with the most profitable results. Thus, the people we manage must be communicated with.

The employee who fails to be adequately involved in an organization's communication is not just missing out on a nice corporate perquisite. In a very real sense, that individual is not a part of the working relationships which constitute the organization. Involvement in communication cannot be viewed as some special gift awarded to employees. It is instead the very process by which people become organized.

How can we avoid sin number seven? Recognize that management *is* communication. Then be willing to spend organizational resources to improve communication quality.

Concerned management can periodically conduct a communication audit—an analysis of the way people are interacting within the organization. Still other companies invest in ongoing communication skills training and retraining to make sure people are consistently aware of the importance of effectively interacting one with another.

The challenge is, then, to sharpen your skills as you repent of the seven deadly sins of management communication.

Reading Comprehension Questions

1. Why does the author want to discuss the seven deadly sins?

2. How does "expecting to be misunderstood" assist communicators?

3. What are the dangers of using impressive-sounding words?

4. How does the medium affect the message?

5. What is the feedback loop?

6. What are some nonverbal ways to communicate negative feelings? What are some nonverbal ways to communicate positive feelings?

7. How can you help the reader or listener get the message?

8. From your experience, describe a manager who communicated either effectively or ineffectively. Explain why.

WRITING: Memos and Forms

Most of the writing required of beginning technicians consists of in-house forms and memos. **Forms** have preprinted formats for recording specific pieces of information. **Memos** (short for memorandums) are brief, open-ended communications addressed to a specific person about a specific subject.

FORMS

The form that is most frequently used by technicians is the *troubleshooting record*. Companies often have their own particular name and format for this type of record. It is sometimes called a *service report, customer service order, trouble log, status report,* or *service repair order.*

Printed on the form will be specific labels, with boxes or lines provided for inserting the information. Common labels are the customer's name (or number), date, time, product name (or number), symptoms (customer's complaint), problem diagnosis, parts replaced (description or part number), the cost of the parts, and the status of the repair (what further work needs to be done). The technician will record information on the form.

The forms may then be used by the company for billing, inventory, and customer account records, and they are usually kept in a central location, often entered into a computer databank. Since many departments may review the record for a variety of reasons, it is vital to record the information neatly and accurately.

Some companies review all records and return incomplete records to the technician for more information. Since the review could take place several days after the service was performed, it would be understandably difficult to recall and supply the missing information.

Troubleshooting procedures may take place over long periods of time. For instance, a replacement part may not be in stock. By the time the part arrives, the original technician may be working on another assignment, and a second technician will have to rely on the first diagnosis. The amount of information that was provided by the first technician will affect how quickly the piece of equipment will be up and running. Companies routinely require progress reports or project updates on unresolved records.

Figure 1 is an example of a service report form. An important item, for the customer as well as the repair company, is the warranty information. This information will affect how the customer is billed and how the factory is billed, and it may even affect the future of the product. If enough products suffer the same defect, it may indicate a factory design problem.

Other common forms are the purchase requisition (Figure 2) and the monthly vehicle report (Figure 3). These forms also ask for specific, precise information that is necessary for efficient accounting, ordering, billing, or reimbursement.

Some points to remember in filling out forms are the following:

1. Write or print neatly. Many people may have to decipher your handwriting. A misunderstanding may have negative consequences for the customer or for you. Correct spelling and clear wording help.

2. Check out any information that you cannot easily supply. This includes the date and time, part numbers or descriptions, and the customer's full name.

3. Do not overlook supplying information you take for granted, such as the customer's complaint. Anticipate the questions that your company may have concerning the service.

4. Record your procedures as soon as possible after you perform them. Some technicians keep a personal work record in addition to the company forms they submit. If questions arise later, or a similar problem comes up, the personal record provides details of past work.

5. Use standard abbreviations and symbols on in-house forms. If the form is given to the customer, use full words. Some customers are reluctant to pay for services they cannot interpret.

Apple Computer, Inc.

Service Repair Order

- PLEASE COMPLETE ALL AREAS -

CUSTOMER INFORMATION	
NAME:	
ADDRESS:	
CITY:	
STATE, ZIP:	
TELEPHONE:	

DEALER INFORMATION

SRO NO. **G 61714**

SPECIAL REFERENCE NUMBER: ______

TODAY'S DATE	PRODUCT REQUIRING SERVICE	SERIAL NUMBER	PURCHASE DATE
/ /			/ /

OTHER EQUIPMENT (IF ANY):

AppleCare® AGREEMENT NUMBER: | EXPIRATION DATE: / /

WARRANTY INFORMATION - Always Check Only One

- ☐ 1 - 30 DAYS
- ☐ 31 - 60 DAYS
- ☐ 61 - 90 DAYS
- ☐ AppleCare®
- ☐ SERVICE WARRANTY REPAIR
- ☐ SERVICE STOCK REPLACEMENT
- ☐ STORE DEMO REPAIR
- ☐ OUT OF WARRANTY
- ☐ OTHER: ______

TROUBLE REPORTED

ESTIMATES | TIME: | COST:

DEALER COMMENTS

QTY.	PART NUMBER	DESCRIPTION	AMOUNT

REPAIRED BY

X

TOTAL PARTS:	
TAX:	
LABOR RATE: ______ HOURS X ______ RATE =	
TOTAL:	

53530626006 1714

DEALER FILE

REORDER INFORMATION (408) 354-7792 - R MILLER

FORM # 073-0101-B

FIGURE 1

No. 793
PURCHASE REQUISITION

Supplier (If Known) _____

Purchase Order No. _____

Date _____ Terms _____

Buyer _____

Non-Taxable ☐ Taxable ☐ _____ %

Ship Via: ☐ Surface ☐ Air ☐ Other

F.O.B. ☐ Destination ☐ Shipping Point

Certificate of Compliance Required ☐ Yes ☐ No

Inspect* ☐ Yes ☐ No

Desired Date	Order Date	Confirmed Delivery Date

Item	Stock No.	Eng Rev	Quantity	Description	Estimated Unit Price	Actual Unit Price	Unit	Var. Code	Gov't Priority Comm Code	Job OR PL-Dept.-Acct.	Government Contract No.

Estimated Extended Value _____   Actual Extended Value _____

NOTE: REQUISITIONER SHOULD FILL IN ALL SHADED AREAS.

Item	Rec'd	Date	Bal. Due	Rec'd	Date	Bal. Due	Rec'd	Date	Bal. Due	Rec'd	Date	Bal. Due
1												
2												
3												
4												
5												
6												
7												
8												
9												

Requested By _____ Ext. # _____ Date _____ Approved By * _____ Date _____

Deliver To _____

For 1635 Electronic Test Equipment

PURCHASING

FIGURE 2

FIGURE 3

MEMOS

A memo may also be a preprinted form, but it is an open-ended form. Some companies supply workers with memos, and other companies expect memos to be typed. Follow your company's policy.

The standard labels of a memo are TO (the reader), FROM (the writer), DATE (day, month, and year), and SUBJECT (the specific topic), which are all followed by colons and the appropriate information.

Memos should address one issue, as the writer indicated in the SUBJECT, and they are usually less than one page long. Although the language may be casual and conversational, it should be written in standard English. Make an effort to communicate completely to prevent misunderstandings.

The following are examples of a poorly written and a well-written memo.

TO: Ed   DATE: Tues.

FROM: Chris

SUBJECT: Scott Van Hoffman

Scott tripped over a toolbox and had to go to the hospital. He was treated in the emergency room and released. Figures he'll be out about 2 weeks. No problem to reschedule.

Poorly written memo

TO: Ed Cramer, Human Resources   DATE: 1/31/87
    cc: Scott Van Hoffman

FROM: Chris Anderson, Foreman

SUBJECT: Incident Report—Scott Van Hoffman
                ID# 30045
                Hire date: 5/15/86

[What happened]
On 1/30/87, 10:15 a.m., Scott Van Hoffman backed up from his workstation and tripped on his toolbox, which had not been returned to its shelf. The accident was apparently caused by his own negligence, but I will gather more information from his co-workers and from Scott by this Friday.

[Result]
I drove Scott immediately to General Hospital, where he was treated in the emergency room (X rays and a prescription for pain) for a sprained wrist and multiple bruises on his tailbone. The doctor released Scott and recommended bedrest for three days, a check-up next Monday, and limited use of his right hand for two weeks. The standard hospital insurance forms have been filled out and submitted.

[Further action]
Scott's absence will require only minor scheduling revisions which are already in process. Scott will be contacting you concerning his disability insurance and workers' compensation benefits.

Well-written memo

Extremely brief memos, such as a question or a simple response to a question, can consist of one sentence. Most memos, however, are longer. Using a written medium for communication makes the message important and permanent. Memos can be used to clarify ideas, state a position, make a request, or supply information. Provide enough discussion to clarify your message, and include the action you are requesting or offering. Try to anticipate questions and provide the information the reader will need to act on your memo.

Some points to remember when writing memos are the following:

1. Single-space memos, and double-space between paragraphs.
2. Use standard spelling, punctuation, and grammar. Write complete sentences.
3. Include the names of those to whom copies of the memo will be sent. They are listed following the abbreviation "cc:" (carbon copy) either in the TO section or at the bottom of the memo.
4. If you type your memo, write your initials beside your name. This serves as a signature.
5. Keep a copy of all the memos you send. Your reply might be a simple "yes," or you may have to follow up on your memo. Always assume that the receiver will keep your memo, too.

Exercise 13.1 Fill out the trouble log according to the following information. Add enough information to make the report realistic.

On 6/10/88 the Acme Component Company called to report a malfunctioning video display terminal. Your contact person is Mr. Washington, manager of the MIS Department. You arrive at ACC on 6/11/88 at 8:15 a.m. to begin troubleshooting. Mr. Washington is in a meeting, but his staff programmer, Jack Haynes, takes you to the VDT. There is a wavy pattern on the screen. You trace the problem to the power supply and replace the rectifier. The VDT then functions properly, and you leave at 9:25 a.m.

TROUBLE LOG

Customer			System Type	Serial No.
	Date	Time		
Problem First Noted			Problem Symptoms (Customer):	
Called				
Arrived				
SVC Complete			☐ System Down ☐ Option Down ☐ Application Down	
Option			LARS No.	OK'd By

Problem Diagnosis
_____
_____
_____
_____
_____
_____
_____
_____

Exercise 13.2 Write a memo to your supervisor, Ms. Franklin. She received a call on 6/12/88 from Mr. Washington, who complained that no one from your company arrived to fix the VDT. A copy of this memo will be sent with the bill to Mr. Washington. Use the standard headings as shown below.

TO: DATE:

FROM:

SUBJECT:

Exercise 13.3 Keep a vehicle report for one week of your travels to and from work and school as though you will be reimbursed for your travel time. Submit the vehicle report below for the final record.

MONTHLY VEHICLE REPORT — _____

EMPLOYEE NAME (PRINT)   COMPANY CAR #

ENDING ODOMETER READING
BEGIN. ODOMETER READING
TOTAL MONTHLY MILEAGE
LESS BUSINESS MILEAGE
TOTAL PERSONAL MILES
RAN # (IF NEEDED)

6
EMPLOYEE #   YEAR/MAKE/MODEL

COST CTR #   OFFICE #   RENTAL CONTRACT #

CASH AND PERSONAL CREDIT CARD PURCHASES ONLY
(ATTACH RECEIPTS FOR ANY EXPENSE OVER $10.00)

DATE	BUSINESS MILEAGE	GALS./LTRS.	GASOLINE $	QTS.	OIL $	ACCOUNT 372 MAINTENANCE & TIRES	ACCOUNT 373 MISCELLANEOUS	TOTAL REIMBURSEMENT

TOTAL

SIGNATURE AND DATE

APPROVAL SIGNATURE AND DATE

TITLE

MILEAGE REIMBURSEMENT: A. FLAT RATE
ACCOUNT 377   B. PER MILE RATE
RENTAL EXPENSE - ACCOUNT 379
GROSS REIMBURSEMENT
PETTY CASH REIMBURSEMENT
NET REIMBURSEMENT DUE

M-1176-1   FLEET - WHITE COPY; ACCOUNTS PAYABLE - CANARY COPY; OFFICE - PINK COPY; DRIVER - GOLD COPY.

164 In-House Communications

Exercise 13.4 Complete a purchase requisition for the parts you need to complete a lab project as though you are ordering the parts from your company's stockroom. Add information to make the requisition complete. Submit the requisition form below.

								No. 793
								PURCHASE REQUISITION

Supplier (If Known) _____

Non-Taxable ☐   Taxable ☐ ___%

Purchase Order No. _____
Date _____ Terms _____
Buyer _____

Desired Date	Order Date	Confirmed Delivery Date	Ship Via: ☐ Surface ☐ Air ☐ Other	F.O.B. ☐ Destination ☐ Shipping Point	Certificate of Compliance Required ☐ Yes ☐ No	Inspect* ☐ Yes ☐ No	Gov't Priority	Government Contract No.

Item	Stock No.	Eng Rev	Quantity	Description	Estimated Unit Price	Actual Unit Price	Unit	Var. Code	Comm Code	Job OR PL-Dept.-Acct.

NOTE: REQUISITIONER SHOULD FILL IN ALL SHADED AREAS.

Estimated Extended Value ___   Actual Extended Value ___

Item	Rec'd	Date	Bal. Due	Rec'd	Date	Bal. Due	Rec'd	Date	Bal. Due	Rec'd	Date	Bal. Due	Rec'd	Date	Bal. Due
1															
2															
3															
4															
5															
6															
7															
8															
9															

Requested By	Ext. #	Date	Approved By	Date
Deliver To			* For 1635 Electronic Test Equipment	

PURCHASING

Exercise 13.5 Complete a business and travel expense report for a business trip. Assume that you have attended a business convention in an American city. Use actual prices, costs, and fares whenever possible. Attach at least one actual receipt that resembles a submitted expense, such as a restaurant, taxi, or parking bill that you have paid. Submit the business and travel expense report on the following page.

BUSINESS AND TRAVEL EXPENSE REPORT ACCOUNTING COPY

NAME (LAST, FIRST, INITIAL)				DEPT. NO.	ACCT. NO.	WEEK ENDING
STREET		CITY	STATE	ZIP CODE	REP # 1-5	REGION 9-12

DAY OF WEEK	MONDAY	TUESDAY	WEDNESDAY	THURSDAY	FRIDAY	SATURDAY	SUNDAY	TOTALS FOR WEEK
DAY OF MONTH (WRITE IN)								
TRAVEL FROM (CITY)								
TRAVEL TO (CITY)								
PURPOSE OF TRAVEL								
BUSINESS MILEAGE								14
MILEAGE ALLOWANCE								01
GAS/OIL								02
PARKING & TOLLS								10
OTHER TRANSPORTATION* (PAID BY EMPLOYEE)								04
MEALS & ENTERTAINMENT*								03
LODGING								08
POSTAL, TELEPHONE								05
MISCELLANEOUS*								09
AIR FARE								
TOTAL (PAID BY EMPLOYEE)								

LESS TRAVEL ADVANCES RECEIVED	DATE		AMOUNT		TOTAL TRAVEL ADVANCES	( )
TOTAL AMOUNT TO BE PAID TO EMPLOYEE (REFUNDED TO DEVRY)						
AIRFARE BILLED TO CO. (ATTACH STUB & INVOICE)						09
APPROVED DIRECT BILLINGS TO CO. (ATTACH RECEIPTS)						

MISCELLANEOUS EXPENSES PAID BY YOU* (ATTACH RECEIPTS)

DATE	EXPLANATION	AMOUNT

OTHER TRANSPORTATION* (ATTACH RECEIPTS)

DATE	EXPLANATION	AMOUNT

DESCRIPTION OF MEALS & ENTERTAINMENT EXPENSE* (ATTACH RECEIPTS)

DATE	NAMES & TITLES OF PERSONS	RESTAURANT/ENTERTAINMENT SITE & CITY	BUSINESS DISCUSSED (IF NOT ALONE)	TYPE OF EXPENSE	AMOUNT

EMPLOYEE SIGNATURE	DATE	A/P CHECK 12	ODOMETER READINGS THIS WEEK	
			ENDING NO.	
APPROVED BY	DATE	AUDITED BY	ADJUSTMENTS 11	BEGINNING NO.
				TOTAL MILES DRIVEN

909235 REV BTER 5510

In-House Communications

Exercise 13.6 Complete an accident report that will be sent to the State Board of Workers' Compensation. Assume that you had an injury at work. Fill in the details and information to make the report realistic. Then type a memo to your supervisor describing the accident. Attach the completed form to your memo. Submit the form below and the memo.

A   **STATE OF GEORGIA**
EMPLOYER'S FIRST REPORT OF INJURY OR OCCUPATIONAL DISEASE

(See Instructions on Reverse Side)

Ga. Form WC-1 (Rev. 7-82)

OSHA File Number

Insurer File Number

DO NOT WRITE IN THIS COLUMN

Employer | Employer Phone No.
Address | Employee Soc. Sec. No.
City | State | Zip | Date of Injury

2. Employee/Claimant Name (Last) (First) (Middle) | 3. Insurer

Address
City | State | Zip

→ COMPLETE ORIGINAL AND ONE COPY AND ←
SEND IMMEDIATELY TO

☐ LUMBERMENS MUT. CAS. CO.   ☐ AMERICAN MOTORISTS INS. CO.
☐ FEDERAL KEMPER INS. CO.   ☐ AMERICAN MFRS. MUT. INS. CO.

**1401 PEACHTREE STREET, N.E.
ATLANTA, GA 30309**

Carrier No.

Sic

4. Nature of Business (Mfg., Trade, Transportation, Etc.) | Specific Products

5. Employee's Home Telephone   Area Code   Number | Marital Status: Single ( ) Married ( ) Divorced ( ) Separated ( ) Widowed ( ) | Date of Birth | Age | Sex: Male ( ) Female ( ) | Age | Sex

6. Regular Occupation | Department in Which Regularly Employed | Occupation

7. Place of Accident or Exposure (Address or Location) | On Employer's Premises? Yes ( ) No ( ) | Location

8. County of Injury | Time of Injury A.M. ( ) P.M. ( ) | Time Workday Began on Day of Injury A.M. ( ) P.M. ( ) | First Date Employer Aware | Employer Aware

9. Describe the injury or occupational disease in detail and indicate the part of body affected. (e.g., amputation of right index finger at second joint; fracture of ribs; lead poisoning; dermatitis of left hand, etc.) | Nature

10. If Fatal: Give Date of Death   Number of Dependents | Body Part

11. What was employee doing when injured? (Be specific. If employee was using tools or equipment or handling materials, name them and tell what employee was doing with them.) | Type

12. How did accident or exposure occur? (Describe fully the events which resulted in injury or occupational disease. Tell what happened and how it happened. Name any objects or substances involved and tell how they were involved. Give full details on all factors which led or contributed to the accident or exposure.)

13. What thing directly injured the employee or made employee ill? (Name object struck against or struck by: vapor, poison, chemical or radiation, if strain or hernia, the thing being lifted, pulled, etc.; if injury resulted solely from bodily motion, the stretching, twisting, etc., which resulted in injury.) | Source

14. Name and Address of Treating Practitioner (Include Zip Code) | Name and Address of Hospital If Hospitalized (Include Zip Code)

15. Did employee work the next day following injury? Yes ( ) No ( ) | First Date Employee Failed to Work A Full Day

16. Time Discontinued Work: A.M. ( ) P.M. ( ) | If Returned to Work, Give Date | Returned at What Wage? $ Per Week

17. Length of Time In Your Employ   years   months | Did employee receive full pay for date of injury? Yes ( ) No ( ) | Wage Rate at Time of Injury or Disease $ | Per hour ( ) Week ( ) | Day ( ) Month ( )

18. Hours Worked Per Day ( ) Week ( ) | Number of Days Worked Per Week | List Normally Scheduled Off Days

19. If employee is paid on commission or piece work basis, enter average weekly amount. $ | If board, lodging or other advantages were furnished, enter average weekly amount. $

20. Report Prepared by (Print or Type Name) | Position | Date of Report

EMPLOYER'S FAILURE TO SUBMIT THIS REPORT TO INSURER IMMEDIATELY MAY RESULT IN PENALTY.

SPELLING/VOCABULARY: Seed Roots

In this chapter we combine the spelling and vocabulary exercises because of the unusual nature of this topic.

The root word we pronounce as "seed" is spelled in three different ways: CEED, CEDE, and SEDE. Each spelling follows certain prefixes, many of which you have already studied.

The reasons for the three spellings are easy: They are derived from two Latin roots with slightly different meanings, and over the centuries, as the words crossed languages, certain spellings for certain words became standard. Knowing this does not make remembering the correct spellings any easier; however, we spell these roots in only a dozen verbs. Related words also use forms of the roots, and knowing the roots gives us clues to the meaning of these words.

SEDE is derived from the Latin word *sedere* meaning "to sit." We use this spelling with only one prefix.

> SUPERSEDE = *super* (over) *sede* (to sit), replace or to cause another to become obsolete
>
> *The transistor supersedes the vacuum tube in solid-state electronics.*
>
> Related words: sedate (to sit calmly)
>
> sedentary (moving little, as in a sedentary lifestyle)
>
> presides (sits before others as head of a meeting)
>
> session (a sitting or assembly of many people)

CEDE is derived from the Latin word *cedere,* meaning "to yield," "to go," or "to leave."

> CEDE = to surrender formally
>
> *Spain ceded Florida to the United States in 1819.*
>
> SECEDE = *se* (apart) *cede* (go), to formally withdraw or separate
>
> *Florida seceded from the Union in 1861.*
>
> Related word: secession (a formal separation)
>
> INTERCEDE = *inter* (between) *cede* (to go), to mediate or make a request on behalf of another
>
> *An attorney can intercede in legal disputes.*
>
> Related words: intercession (the mediation or pleading for another)
>
> intercessor (a person who intercedes)
>
> ANTECEDE = *ante* (before) *cede* (to go), to go before in rank, time, or place
>
> *The vacuum tube anteceded the transistor.*
>
> Related words: antecedent (going before in time, logic, or order, prior)
>
> ancestor (one who lived earlier in a family line)
>
> PRECEDE = *pre* (before) *cede* (to go), to go before in rank, time, place, or importance

The development of solid-state electronics preceded the use of integrated circuits.

Related word: precedent (a fact or procedure established before)

CONCEDE = *con* (with) *cede* (to go), to admit as true or acknowledge

After the votes were counted, the loser conceded the election to his opponent.

Related word: concession (a privilege, right, or lease)

RECEDE = *re* (back) *cede* (to go), to go or move back

After the floodwaters receded, people returned to their homes.

Related words: recess (a temporary halt or withdrawal)

recession (a departing processional or inactivity in the economy)

recessive (a nondominant gene)

ACCEDE = *ac* (to) *cede* (to yield), to consent or enter upon the duties—rarely used except in sophisticated, formal language.

Both nations acceded to the treaty.

Related words: access (approach or come near)

accessory (helping in a subordinate way)

accessible (can be approached easily)

CEED is also derived from the Latin root *cedere,* meaning "to go." This spelling is used with only three prefixes.

EXCEED = *ex* (beyond) *ceed* (to go), to surpass or outdo

The driver was fined for exceeding the speed limit.

Related words: exceeding (extraordinary)

exceedingly (extremely)

excess (an amount more than is needed)

excessive (being too much or too great)

PROCEED = *pro* (forward) *ceed* (to go), to advance, or to go on after stopping

After a brief recess, the lawyers were asked to proceed with the trial.

Related words: process (a method of development)

procession (moving forward, a parade)

procedure (a step in a process)

procedural (having to do with a step)

SUCCEED = *suc* (under) *ceed* (to go), follow or come after; to happen or turn out as planned; have a favorable outcome.

The election will determine who will succeed the outgoing president.

Related words: succession (series or sequence)

success (favorable outcome or result)

successful (having achieved success)

SPELLING REVIEW

SEDE	CEDE	CEED
supersede	accede	exceed
	antecede	proceed
	cede	succeed
	concede	
	intercede	
	precede	
	recede	
	secede	

Remembering the correct spelling for these words will take some memorization and frequent use. Use the dictionary when you are in doubt. Since there is only one word using SEDE, you should not have too much trouble remembering it.

A crutch for remembering the three words using CEED is to think of the word SPEED. Speed ends with the EED ending, and the first letters of *succeed, proceed,* and *exceed* form the first three letters of SPEED. All the other words use the CEDE spelling.

Exercise 13.7 Fill in the missing letters to form words derived from SEDE, CEDE, or CEED roots.

1. The pro_____dures of the meeting were determined by the presider.
2. The issue of retroactive pay raises super_____ded all other issues.
3. The presider allowed arguments concerning the high costs of living to pre_____de other items on the agenda.
4. The economy was not expected to re_____de in the suc_____ing months, according to the market forecaster.
5. Inflation was making meeting day-to-day expenses ex_____dingly difficult.
6. The accountant was asked to inter_____de for the employees at the budget negotiations meetings.
7. The accountant was given ac_____ss to all company financial records.
8. The CEO con_____ded that harmony within the company was the primary goal of management.
9. As the negotiations pro_____ded, the workers became more optimistic about a suc_____ful outcome.
10. The day the new policy went into effect, a valuable pre_____dent was established which ex_____ded the employees' expectations.

VOCABULARY REVIEW

Exercise 13.8 Use the dictionary to find *two* other words using each prefix. Write each word and its meaning.

1. Ante- (meaning "before") _____

2. Pre- (meaning "before") _____

3. Pro- (meaning "forward") _____

4. Con- (meaning "with") _____

5. Re- (meaning "back") _____

6. Ex- (meaning "beyond") _____

7. Ac- (meaning "to"; could be spelled ad-) _____

8. Super- (meaning "over") _____

9. Suc- (meaning "under"; could be spelled sub-) _____

10. Inter- (meaning "between") _____

WORD WATCH: Past/Passed

The words PAST and PASSED are pronounced the same. The simplest uses of these words are easily categorized.

PASSED is a verb, the past tense of PASS. Use PASSED if the word is a verb or part of the verb in the sentence.

> He passed the test.
> She has passed back the papers.

PAST can be used as a noun, adjective, adverb, or preposition. Use PAST for any use other than a verb.

> His past employers speak highly of him. *(Adjective)*
> In the past, he has worked hard. *(Noun)*
> I wouldn't put it past him to work overtime. *(Adverb)*
> He has gotten past the first interview. *(Preposition)*

Exercise 13.9 Write the correct form of PASSED/PAST for each sentence.

1. Customers who have had poor service in the _____ are likely to tell their friends about it.
2. Customers who received good service for a fair price may not have _____ the news to others, but tend to stay customers.
3. Bad news gets _____ on faster than good news.
4. Customers who have had good service and friendly service in the _____ usually remain customers.
5. It is twice as expensive to recruit new customers as it is to hold _____ customers.
6. Companies with cold representatives and uncaring technical staff will often be _____ over for a friendlier company.

7. The business that _____ on helpful information to the customer in the _____ was remembered.
8. The field technician is usually the representative closest to the customer once the customer is _____ the sales calls.
9. Field technicians who are friendly, competent, and efficient are valued by their company for retaining _____ customers.
10. The technician who couldn't relate to his customers, even though he was technically competent, was _____ over for promotion.

CHAPTER FOURTEEN

BUSINESS LETTERS

- *Correct ineffective wording.*
- *Type a correct, formal business letter and envelope.*
- *Type a formal letter requesting information.*
- *Type a letter to an elected official.*
- *Type a letter of goodwill: sympathy, congratulations, or thanks.*
- *Type a negative letter refusing information.*
- *Spell* ible/able *endings correctly.*
- *Use* grad/gress *roots correctly.*
- *Use* stationary/stationery *and* compliment/complement *correctly.*

READING: Turning Confrontation into Communication

In today's competitive world, confrontation is inevitable. The possibility of a faceoff exists in every facet of your life—professional as well as personal.

It exists at public meetings and in private arguments, at the office and at home, with your senior partner and with your son, in practically every situation involving more than one person. You cannot pick the time or the place for a confrontation, but you can control your response and even the terms of the debate—and thus convey your ideas effectively.

First, concentrate on your attitude. When someone hits you, you want to strike back. When someone shouts at you, you want to shout back. Though a natural tendency, it is not a winning one. In fact, it is a waste of energy. Do not shout denials. Calmly explain your stand and be positive.

Next, concentrate on breathing and relaxation. Proper breathing creates an almost instant feeling of well-being. You will find you can immediately banish tension and stress. The relaxed, self-assured person can think—and the thinking person can take control.

The key to proper breathing involves moving the diaphragm (the muscle just below the rib cage) to make room for your lungs to fill with air. Difficulty arises when we incorrectly move the diaphragm *in* on inhalation and *out* on exhalation. We fall into that bad habit when we are nervous, thus reducing our oxygen supply just when we need it most. The trained singer practices breathing exercises as assiduously as scales. You must do the same.

Next, know exactly what you want to do. If you are certain that a meeting has potential for confrontation, take time to plan your strategy. Determine what you want

ARCH LUSTBERG

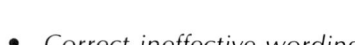

Disputes are inevitable. The first step toward winning: Control yourself.

"Turning Confrontation into Communication," by Arch Lustberg. Reprinted by permission from *Nation's Business,* June 1984. Copyright © 1984, U.S. Chamber of Commerce.

to accomplish and how you can best make your point. Never lose sight of your objective.

Do not spoil your planning by getting irritated and snapping at the person you are talking to. You will undoubtedly say something you will regret. And if that happens when you are talking with the media, your unfortunate comment may be used to conclude a broadcast interview—or it may be the only thing you say that is broadcast.

Expert communicators direct their thoughts to those people who do not have an opinion on the subject at issue. (This "audience" may even be imaginary, as when you and your adversary are the only people present at a meeting.) An upbeat statement directed toward them will serve your purposes much better than a defensive quip aimed at your adversary.

Speak simply, clearly, concisely. We have become a people who think that we have to use big words or jargon to appear "in the know." I recommend that you replicate, interface and offload only in the privacy of your own home. Speak English in public.

All of us have four weapons that will help us communicate: the mind, the face, the body and the voice. Though we generally use them correctly in animated conversation, we almost never use them correctly when we are tense, afraid and intimidated.

How can you use your mind creatively when face to face with an adversary? The best tool is the pause. It gives you time to think of a positive response, time to eliminate negative comments.

Remember, however, that the pause will work only if it looks comfortable. The key is to remain silent and maintain eye contact with the person you are talking to.

This means that you should avoid such audible pauses as "uh ... uh ... ," "like," "I mean" and "you know" and that you should not give the impression that you are afraid to look your adversary in the eye. Eye contact does not necessarily mean eye to eye. If you find that uncomfortable, try focusing on a certain part of the person's face. Most people, in fact, look at the lips.

Television provides a wonderful opportunity to study the pause. The next time you are watching an interview, notice how the pause—when used correctly—makes a person appear more confident.

What about effective use of the face? The smiling, animated face is one of your strongest communications tools. And it is one you will probably use most often after you have mastered certain basics.

All of us usually "open" our faces when we are talking with someone. The problem is that in confrontational situations we often close the facial muscles, creating a frown line that we think makes us look "professional."

Try this excercise in front of the mirror to get your facial muscles to work for you: First, tighten the brow as much as you can. Hold that position and count aloud from 1 to 5, then relax. Repeat the exercise, counting a bit louder. Now open your brow. Make the lines in your forehead that come when your eyes are wide open and your eyebrows are arched. Hold the position and count aloud to 5.

Repeat the exercise. This time count as quietly as possible. You are not only opening your face, you are also speaking in a warmer, friendlier, more communicative way.

Get your body into the act of communicating, too. Probably the most animated forms of communication are gossiping and telling secrets. Watch two people deep in conversation. They are not just talking, they are painting pictures with their faces. Their entire bodies are alive.

You can do the same. Talk with your body as well as with your face.

The hands are the most used—and abused—part of the body when it comes to communication. You need to make them work in concert with the face—and make them less conspicuous when that is appropriate.

All of us tend to hide our hands, just as we do our faces. You probably clutch one in the other, hold them behind your back, put them in the folds of your arms or put one in a pocket and grab your wrist or forearm with the other. You thus end up calling attention to your hands—not communicating.

Try standing with your feet about as far apart as the width of your shoulders. Shake your shoulders a few times. Then let your hands fall naturally. Now you are in position to use your hands in animated conversation. When you begin talking with someone, keep your hands and fingers still. Make gestures when you feel they will complement the points you are making with your face and voice.

Your voice tends to follow the personality created by your face and body. When one is warm and friendly, so is the other. When one is cold and hostile, the other follows suit.

You can add interest and drama to your voice by learning when and how to use pitch, rate and volume. Pitch is the position of the sound on the musical scale. Rate is the duration of sound. Volume is the decibel level.

Volume is the least effective vocal tool. It is useful only when your purpose is to discipline a child or a pet. Loud sounds are irritating.

To communicate, you must express yourself. The best expression comes with uninhibited and unselfconscious use of pitch and rate. For example:

174 Business Letters

- "I had a wonderful time." (Unless you do someting wonderful with *wonderful*, your host will think you are lying.)
- "It was a magnificent day." (Make *magnificent* truly magnificent.)
- "That garbage gave off the most foul smell." (It couldn't have been worse!)
- "Give me that knife." (If you don't, you'll be sorry.)

That is vocal flexibility: the willingness and ability to use these vocal tools.

So there you have the steps you can take to become an effective communicator—to win at confrontational situations. You can adapt them to any situation. If you are prepared, you will have an excellent opportunity to win.

You have something valuable to say. Learn to say it well.

Fear of Speaking

For thousands of Americans, fear of public speaking has serious consequences because their jobs require them to give testimony before local, state or federal legislators.

Their fear—combined with a lack of the proper communications skills—often keeps them from making the impact they want.

Among the types guaranteed to turn off an audience:
- The droner who delivers a boring text without vocal inflection, facial expression or eye contact.
- The ill-prepared scaredy-cat who stumbles through his text with an assortment of nervous tics.
- The long-winded bore who thinks more is better.
- The greeting reader who keeps his eyes glued to the page, even reading "good morning."
- The statistician who overwhelms his audience with numbers that no one can relate to.

You can avoid falling into those deadly categories by mastering some basic principles:
- Strive for an easy, open facial expression.
- Use body language to emphasize words.
- Pause to allow time to breathe and throw off stress.
- Control your voice, varying your inflection and emphasis.
- Maintain eye contact with your audience.

You can also make your testimony more effective by keeping it simple and short.

Remember, you want to communicate ideas, not just read words. Testimony that is well delivered will win the day.

Adapted from Testifying With Impact, *by Arch Lustberg.* © *1983 by the Association Department, U.S. Chamber of Commerce.*

Reading Comprehension Questions

1. Define a confrontation as the author uses the word.

2. What advantage does remaining calm give you in a confrontive situation?

3. How can you use your mind as a "weapon" in a confrontive situation?

4. How can you use your face and body as weapons in a confrontive situation?

5. How can you use your voice to gain control of a confrontive situation?

6. Describe an experience of yours in which you handled a confrontive situation maturely and effectively.

7. From the "Fear of Speaking" section, describe one basic principle of good speaking that you need to work on.

WRITING: Business Letters

Letters usually fall into two groups: personal and business. Personal letters can be written in informal, breezy styles and are often written just as the writer would speak—with slang, familiar expressions, and a variety of loosely organized topics.

Business letters are written in formal language and standard styles. Like all technical writing, business letters should be clear, concise, and to the point. Letters may be longer than one page and cover more than one topic, but the ideas and paragraphs are organized and presented in a thoughtful, deliberate way. Business letters are formatted in either the "blocked" or "semiblocked" style. The examples demonstrate these formats.

Many people are reluctant to write business letters because, despite their original purpose, the words used in a letter can portray the writer as either mature or childish, sincere or phony, direct or evasive. The words convey the attitude and image of the writer. Also, letters are often the first and last contact with customers and business associates, which means that they provide the first and last impressions of the writer.

Fortunately, with practice, letters become easier to write. Experience with writing different types of letters now will make your future letter writing less stressful and less time consuming.

In this chapter you will see and practice several types of letter. Practice using both formats of letters.

BLOCKED AND SEMIBLOCKED FORMATS

The two standard formats for business letters are the blocked (Examples 1 and 3) and semiblocked (Examples 2 and 4) styles. Although they are similar in many ways, the two formats vary in the placement of the writer's address and the closing.

The **blocked style** is the simplest: Start every part of the letter at the right margin, skipping lines between the parts and paragraphs as indicated above. Do not indent paragraphs.

In the **semiblocked** or **modified style,** indent the writer's address (unless using letterhead stationery), the date, and the closing and signature to a consistent column number, usually slightly to the right of the center. You may indent each paragraph a consistent number of spaces (5 to 10) or not indent any paragraphs. Choose one method and be consistent.

General Letter-Writing Hints

- Always type business letters.
- Single-space letters, but skip lines between the parts and paragraphs of a letter.
- Use the margins (top, bottom, right, and left) to create a "frame" around the words, and pay attention to the appearance of the letter.
- If a letter is more than one page, number the pages, starting with page 2, in a consistent place (upper right corner or bottom center).
- Use envelopes that are the same width as the stationery.

```
              CIRCUIT MANUFACTURING INCORPORATED
                      1800 OVERLOOK LANE
                   BISMARCK, NORTH DAKOTA 58501
```

April 15, 1987

Short Circuit Associates, Inc.
2110 Hotwire Drive
Minneapolis, MN 55455

Dear President:

As the accounting manger of CMI, I am responsible for keeping our company's computerized accounting and scheduling functions competitive in an aggressive electronics industry. Recently we have become dissatisfied with our computer's inability to handle our growing business.

Specifically, our current software has not been able to allow us the scheduling flexibility we require. We have also been unable to expand our inventory program to accommodate higher volumes. Our IBM System/36 seems slow and sluggish, which we realize is due to outdated software. I have heard about your PACS S/36 package and am interested in discussing its potential to handle our needs.

Please have a representative call me early next week. I am available any weekday, and I would prefer to receive your call after 1 p.m.

Cordially,

Alexander B. Hay

Alexander B. Hay
Accounting Manager and Vice President

ABH: lan

Envelope:

```
A.B. Hay                                              ***Stamp***
Circuit Manufacturing, Inc.
1800 Overlook Lane
Bismarck, ND 58501


                                    President
                                    Short Circuit Associates, Inc.
                                    2110 Hotwire Drive
                                    Minneapolis, MN 55455
```

EXAMPLE 1   Letter of inquiry; blocked style.

SHORT CIRCUIT ASSOCIATES, INC.
2110 Hotwire Drive
Minneapolis, Minnesota 55455
April 20, 1987

Mr. Alexander Hay, Accounting Manager
Circuit Manufacturing, Inc.
1800 Overlook Lane
Bismarck, ND 58501

Dear Mr. Hay:

It was a pleasure talking with you today. Based on our discussions, I believe that our manufacturing software may have the solution for your company's needs.

As the next step, I've enclosed a S/36 Overview brochure, which will give you more detail on our products and our company. As I indicated on the phone, our company understands your company's requirements:

1. The software has to have the capacity to produce the master schedule.

2. Deviations from plan in the manufacturing process have to be reflected back into the master schedule.

3. Inventory counts have to be accurate.

4. Bills of material have to be accurate.

5. Vendor/customer relations must be friendly.

6. Education must be provided to the users.

One of the items in the brochure that I would like to highlight is our manufacturing background and expertise. It takes people who have been on the production line to understand manufacturing and assist a company to achieve its objectives. We have the ability to see the overall process and understand your environment.

I'll call you next week, after you have had a chance to review the material, to set up a date for a product demonstration at CMI. If you have any questions before then, please feel free to call me.

Sincerely,

Charles R. Thomas

Charles R. Thomas
President

CRT:ajr

Enc: PACS S/36 Overview brochure

EXAMPLE 2   Follow-up letter; semiblocked style, indented.

PARTS OF A LETTER

The main parts of a letter include the following:

1. Writer's address
2. Date
3. Reader's address
4. Salutation or greeting
5. Body of the letter
6. Complimentary closing
7. Signature and name of writer
8. Typing information
9. Copy information
10. Enclosure and attachment information

1. *Writer's address.* Write your complete mailing address. Align the right margins of each line either at the right margin (blocked format) or at a midpoint on the line (semiblocked format). This part is unnecessary if you are using letterhead stationery. Include the following:

- Street address, apartment number, or office number
- City, state, and zip code

2. *Date.* Write the date of the letter directly below your address or one line below the letterhead. Use the American style (August 15, 1988) or the international style (15 August 1988).

3. *Reader's address.* Skip two lines. Align the reader's exact address with the right margin of the letter. Include the following:

- Full name
- Title, department (if important)
- Company name
- Street address, suite number, or P.O. box number
- City, state, and zip code

Take care to spell the person's name and the company name correctly—do not guess. If the person has a significant title of one or two words ("President"), type the title after the name. Type titles of more than two words ("Research and Development Manager") on the next line. Use the phone book or a business card or make a phone call to verify spellings and titles.

4. *Salutation or greeting.* Skip two lines from the address. Start the greeting at the right margin. The most common greeting is "Dear" followed by the formal name and a colon ("Dear Ms. James:"). If the person has a distinctive title, address the person by the title ("Dear Dr. Potter:"). Do not use a first name unless the person is a friend or close business associate, and then follow it with a comma ("Dear Bob,").

In ambiguous situations, those in which you do not know who will be reading your letter, use a nonsexist, general greeting such as "Dear Sir or Madam" or "Dear Personnel Manager." Reserve "To whom it may concern" for impersonal situations, not when you expect a positive response or personal attention.

5. *Body of the letter.* Skip one line from the greeting. Skip one line between each paragraph.

The three main parts of the body (often comprising three paragraphs) are the introduction, which states the purpose of the letter, the middle, which includes the information and details pertaining to the purpose, and the conclusion, which states what action would be necessary to resolve the purpose and closes the letter.

Letters generally are written for two purposes: to make a request (for information, material, approval, or a job) or to provide information (directions, project status, or incidents). The tone of letters can be positive (making or responding to requests) or negative (declining requests).

Politeness and courtesy are crucial in the business world, even in negative letters. Avoid all slang; write in complete, coherent sentences; and use correct spelling, grammar, and punctuation. Above all, write in clear language—remember that your reader can read only your words, not your mind.

---------------- **WARNING** ----------------

Language that sounds flowery ("My very dear friend") detracts from the seriousness of a business letter. Be sincere but not gushy.

Language that sounds artificial ("at this point in time") may get an equally artificial response, or no response at all. Be human.

Language that is too technical or too simple is insulting. Be appropriate for your reader.

6. *Complimentary closing.* Skip one line from the last line of the body. Capitalize only the first word of the closing. Several common expressions for closings are the following:

Sincerely,	Cordially,
Sincerely yours,	Cordially yours,
Respectfully,	Yours truly,

7. *Signature and name.* Skip four lines to leave room for a handwritten signature. Then type your full name (usually first name, middle initial, and last name—no nicknames) lined up with the closing. Write your title (if significant) directly below your name. Sign your name in ink between the closing and your typed name.

Once you sign the letter, you are responsible for what it says, no matter who types it. Proofread your letters carefully.

The last three parts of a business letter are included ony if the letter requires them.

8. *Typing information.* If someone else did the typing of your letter, indicate the writer and the typist by initials at the lower left of the letter. Skip one line after the name, and type your (the writer's) initials in capital, unspaced letters followed by a colon, and the typist's initials in unspaced lower case letters (AR:mo). Do not place periods after the initials.

9. *Copy information.* If copies of your letter are being sent to people other than the person who is being addressed (such as your supervisor or the person's supervisor), skip one line from the typing information and type "cc:" followed by the names of those receiving copies of the letter.

10. *Enclosure or attachment information.* When you place something else in the envelope with the letter, let the reader know. Skip one line from the copy information. In the lower, left corner type "Attachment:" (for material stapled or clipped to the letter) or "Enclosure:" (for unattached materials) followed by a one- or two-word description of the material.

Enclosure: Sales brochures

ENVELOPES

Use envelopes that match the stationery, if possible. Always include your name and return address in the upper left corner of the envelope. If your envelope has a preprinted company address, type your name just above the printed return address.

Type the address of the person or company to whom you are writing in the center of the envelope. Single-space addresses of more than three lines. Single- or double-space addresses of three lines. Always verify addresses before mailing to avoid delivery delays. When writing to a specific person, that person's name (and title) should be the first line of the address. If the specific person is unknown, either write the department name as the first line ("Shipping Department") or write an "Attn:" (for attention) and the department name at the bottom of the envelope. Remember to include all known address information, including a P.O. box, suite, or office number.

TYPES OF LETTERS

Letters are either informative, persuasive, or a combination of both. They can be positive or negative in tone, depending on your purpose. They can be formal or informal, depending on how well you know the reader.

Technicians may be expected to write several types of letters, such as a request for sales or technical information from a manufacturer, a price quotation to a customer, a letter of acknowledgment for good service, or a letter of complaint about poor service.

Effective letters must be written in simple, clear language and must include accurate facts. An incorrect part number or a poorly worded request may lead to confusion or errors.

QUESTIONS TO ADD FOCUS AND PURPOSE

Some questions to ask yourself before beginning a letter are the following:

- What results do I want from this letter?
- Who is my reader?
- How familiar is my reader with the topic?
- What does my reader need to know?
- What do I need to know?
- What relationship do I want with the reader?
- How do I expect the reader to react to what I am saying and how I am saying it?
- What impression am I making on the reader?

The first question is the most basic, and without a clear answer, it is doubtful that the letter will accomplish any purpose. When you know what you want, you can communicate it. Knowing the reader will make some word choices easier. Knowing your purpose will help identify appropriate and important information to request or provide. Leaving out relevant information will delay results. The last three questions will affect the tone of the letter: how formal, how friendly, and how direct to write the letter.

FORBIDDEN WORDS AND PHRASES

Several words and phrases are overused, evasive, redundant, or wordy. Avoid them in any kind of writing, particularly business letters.

Wordy phrases:

as I am sure you know	I would hope
as of this date	I would like to express
as you are aware	I share your concern
as you know	it is my intention
at this point in time	it is our un/pleasant duty
at the present time	we regret to inform you
	we are cognizant of the fact

Writing

Overused words and phrases:

access	meanwhile
best wishes	more importantly
bottom line	mutually beneficial
delighted	needless to say
different from	ongoing
enclosed herewith	orient
finalize	parameter
glad	personally reviewed
great majority	prior to
happy	prioritize
herein	respectfully requested (submitted)
hereinafter	share
hopefully	specificity
however	subject matter
image	subject to your approval
inappropriate	therein
input	thrust
in the amount of	to impact
in the near future	to optimize
institutionalize	untimely death
interface	utilize
kindly favor us	very much
maximize	viable

Redundancies:

important esssentials	serious crisis
final outcome	personally reviewed
end result	my own . . .
new initiatives	I, myself, . . .

Use clear, direct language, free of wordy and evasive phrases, to make letters effective and easy to understand.

Exericse 14.1 Rewrite the following sentences in clear language that avoids the forbidden words and phrases.

Example: The end result of the new initiatives will impact your department.
Rewritten: The results of these changes will affect your department.

1. From this point in time, this facility will utilize Ace Trucking for all ongoing shipments.
2. The great majority of personnel respectfully request your immediate attention to the following subject matter.
3. Kindly favor us with your input on this issue so that we can provide viable alternatives.
4. We regret to inform you of the untimely death of your colleague.

5. I would very much like to interface with you and finalize our plans.
6. Please submit payment in the amount of $150.
7. Needless to say, the end results will be mutually beneficial.
8. We would like to express best wishes for your achievements and hope that we can utilize your services in the near future.
9. The research document, enclosed herein, contains the important essentials of the product's ongoing development.
10. I have personally reviewed the plans and have decided to impact our image by maximizing our recent achievements.

Exercise 14.2 Write a formal one-paragraph letter informing your employer of each of the following items:

- a permanent address change
- a two-week vacation date
- a two-week termination notice

Exericse 14.3 Type a formal letter including all ten parts and the envelope for one of the items in the exercise above. Type the letter in either the blocked or the semiblocked format.

Exercise 14.4 Type a formal letter to a company to request information on one of its products. Type the letter in either the blocked or the semiblocked format.

POSITIVE LETTERS

A positive letter expresses information in a neutral or positive tone. The writer could be making a request, providing information, or acknowledging good service. Some letters can be extremely brief—one or two sentences.

> Please send me your current catalog and price list of electronic components. Thank you.

Longer letters are made up of three distinct parts, usually presented in three or more paragraphs.

1. Introduction
2. Explanation
3. Resolution and closing

The introduction states the subject clearly and directly. It may also include the purpose for writing.

> You requested an itemized proposal for the installation of the IBM CAD/CAM system. The following list includes the specific modules, their costs, and the projected installation dates.

The middle part of a letter explains the request or provides the information. Usually, the information is organized and presented in an order that is easy to read and understand. Remember that readers sometimes skim letters, or read them quickly. Complicated material must be explained as simply and accurately as possible. Some explanations contain specific points that need special attention. These points can be highlighted by individual paragraphs, numbering, or listing. Skipping lines between numbered or listed items adds visual emphasis.

The new location is desirable for the following reasons:
1. It has easy access to major traffic routes.
2. It is close to leading customers.
3. It includes reasonable land and construction costs.
4. Part-time employees are available from a nearby technical institute.

The resolution paragraph states the specific action that will accomplish your purpose for writing. It could be a request, date, or statement of commitment. State the specific response that you expect from your letter. Provide exact instructions or requests. Then close the letter with a statement of appreciation for the reader's attention, time, or interest.

I will inform you of our decision by January 31. Thank you for your interest in our plans.

Sometimes no resolution is required, and the writer can simply close the letter. Do not extend the letter with excessive flowery language, but conclude the letter with a positive comment—either appreciation or acknowledgment.

Congratulations for your fine work.

Letters of "goodwill" are those expressing compassion—sympathy, appreciation, or congratulations—to business relations or associates. These letters must be written carefully and simply.

---

600 Cornwallis Court
Roanoke, Virginia 24005
May 19, 1987

Ms. Jody Longley
3605 Winning Way
Seattle, Washington 98105

Dear Ms. Longley:

Congratulations on your announced promotion. After all your hard work and determination, you deserve the recognition. For a change, we have an effective computerized production system in the Seattle plant. I cannot help but think that the successful implementation of the three Puma robots was also due to your commitment and expertise.

I am very happy for you and proud to be working for a company that rewards and promotes employees of your caliber.

Regards,

Calvin M. Shargon

Calvin M. Shargon

EXAMPLE 3 Letter of goodwill; blocked style.

People appreciate brief expressions of goodwill, but they may not appreciate a long, detailed account of your similar experiences and how you handled them. Try to keep messages of sympathy positive by not dwelling on the loss, but rather, by mentioning something positive (such as a pleasant memory) about the person. Letters of thanks can recount exactly the services you received from the reader, as well as an offer to return the favor. Letters of congratulations offer sincere praise for an achievement, without excessive praise that sounds phony. Avoid mechanical language in any of these letters.

Exercise 14.5 Write a letter to an elected official (local, state, or national) informing the official of your opinion on a political issue. Begin by answering the "Questions To Add Focus and Purpose" listed in this chapter. Gather the information you need to write an informed letter. Write the letter in three parts: the introduction, body, and conclusion. Include all the standard parts of a letter. Proofread and type the letter and envelope.

Exercise 14.6 Type a letter to a business associate for each of the following occasions:

- the death of a family member
- her promotion to vice-president of a company
- thanks for referring a customer to you

NEGATIVE LETTERS

Letters are often written to refuse service or employment, to turn down claims or credit, or to deny information. Since the reader will find the news disappointing, it is important not to confuse or insult the reader and make further association with the reader more uncomfortable than necessary. It is possible to continue a business relationship despite this negative situation.

These letters usually have three parts:

1. Introduction
2. Reasons for refusal
3. Refusal and closing

This introduction states the purpose for writing. It recounts the incidents that occurred to initiate the letter. Write the introduction objectively, stating facts that both you and the reader accept as facts. Avoid accusing language such as "you claim" or "you said that." Avoid any wording that is judgmental or belittling, such as "I could have expected this of you" or "How could you?" Accurately state the facts, as in the example below. In this example the writer is refusing the reader's request to redirect a service bill to the factory.

> I serviced an Apple IIc on March 19 at your Riverbend office complex. I replaced the motherboard and cleaned the drive heads. I tested the computer with your software and did not see any further problems. I left a bill for these services totaling $200.

The reasons for refusal are stated to clarify your position and reassure the reader that you considered the request fairly and can justify your refusal. Often, company policies or procedures can be cited which then provide guidance for the reader's actions in the future. Reasons for refusal should be consistent and legally sound. Some companies provide legal advice for these types of letters. Avoid sounding evasive or as though you are blaming the refusal on someone else, such as "this was the manager's decision, not mine." If you are unclear of the reasons, find out the reasons before you begin writing, and write with confidence and conviction.

> The Apple IIc that I worked on was out of warranty. The problems in the CPU and disk drive were caused by normal wear after heavy regular use. I found no evidence of equipment defects or malfunctions in either the CPU or the disk drive.

The final paragraph should clearly state what is being refused, and what, if any, further action can be taken. Acknowledging the reader's point of view, without changing your position, will make you sound more understanding and, thereby, make your decision more tolerable. Close with a positive, sincere statement that will keep the door open for further business or association with the reader.

Universal Communications Company
1 Communications Plaza
Fort Walton Beach, Florida 32549

March 4, 1988

Mr. Carlos Torres
Graduate Placement Office
South Florida Institute of Technology
Miami, Florida 33157

Dear Mr. Torres:

I received your résumé and application for a service technician position at Universal Communications Company. Your education appears to cover the fundamentals of electronics and electricity. Your GPA is admirable and shows that you have learned the basics well.

The company is presently looking for technicians with experience in maintaining digital devices, an area that does not appear on your résumé. Over half of the 600 applicants for the current positions already have this experience.

Because of this missing qualification, Mr. Torres, I am unable to consider you for our present positions. I will, however, place your application and résumé in our files for six months in case a position opens for which you are qualified. I thank you for applying with UniCom. Your qualifications look promising for a successful career in electronics. Best wishes for the future.

Sincerely,

Tracey R. Canfield
Tracy R. Canfield
Human Resource Manager

TRC:na

EXAMPLE 4 Negative letter; semiblocked style, not indented

Your claim for factory reimbursement has been denied and payment for my services is expected from you by the 15th of this month. Be assured that our company does not bill owners for factory defects.

Our company has always strived to give prompt, dependable service at reasonable rates. We value you as a customer and hope that you will continue to use our company to maintain your electronic equipment.

Remember that a firm "no" is easier to handle and tolerate than a runaround or foggy response that leaves the issue unresolved. Also, remember that letters written in an emotional state usually reflect that emotion. Writing letters in a biting, angry tone (no matter how justified) does not make good business sense since they produce negative results. Letters that are objective and rational leave readers their dignity and allow communication and business to continue.

Exercise 14.7 Type a formal, negative letter for the following situation.

You work for a large company that manufactures a highly competitive product. A competitor has written a letter to you requesting some design specifications about the product. You have met this competitor on several occasions, but you are not personal friends. You know that it is against company policy to discuss the product design specifications.

Supply any missing information.

SPELLING: IBLE/ABLE Endings

We add -IBLE or -ABLE to turn words into adjectives. The choice between spelling the ending IBLE or ABLE at the end of a word is usually determined by the Latin root word, as we found for the endings -ANCE and -ENCE. Few people these days are familiar with Latin, but we have some general guidelines to help make this spelling problem easier. Remember that a dictionary is always the best consultant since even the following guidelines have many exceptions.

Use -ABLE with the following groups of words:

1. Most full English words (exceptions listed under -IBLE):

    available   correctable
    dependable   predictable

2. Full English words that had a final dropped E:

    excitable   presumable
    desirable   usable
    flammable

3. Roots that end in a final I:

    appreciable   satisfiable
    justifiable   sociable

4. Words that use the letter A in other endings:

>   demonstrable   (demonstrate)
>   hospitable   (hospitality)
>   inseparable   (separate)

5. Roots that end with a hard C (pronounced K as in *cat*) or a hard G (pronounced G as in *get*):

>   educable   navigable
>   practicable   indefatigable

Some common exceptions require keeping the final -E:

>   replaceable   manageable
>   traceable   salvageable
>   serviceable   changeable
>   noticeable   chargeable

Use -IBLE with the following groups of words:

1. Roots that are not full words.

>   audible   possible
>   invisible   terrible
>   indelible   feasible

2. Some full English words to which an immediate -ION could be added:

>   collectible   (collection)
>   connectible   (connection)
>   reversible   (reversion)
>   suggestible   (suggestion)
>   combustible   (combustion)
>   diffusible   (diffusion)

3. Roots that end in -NS, -SS, or -MISS:

>   defensible   dismissible
>   sensible   permissible
>   accessible   admissible
>   depressible

4. Roots that end in a soft C (pronounced S as in *cent*) or a soft G (pronounced J as in *general*):

>   forcible   tangible
>   deductible   illegible
>   reproducible   intelligible

5. Exceptions to ABLE spellings:

>   flexible   resistible
>   collapsible   discernible
>   fusible

Exercise 14.8 Practice using ABLE/IBLE correctly by adding the ending to each root, and then by spelling the entire word.

ABLE		IBLE	
avail _____	_____	aud _____	_____
depend _____	_____	tang _____	_____
excit _____	_____	invis _____	_____
us _____	_____	feas _____	_____
appreci _____	_____	collect _____	_____
soci _____	_____	revers _____	_____
demonstr _____	_____	suggest _____	_____
practic _____	_____	flex _____	_____
service _____	_____	collaps _____	_____
notice _____	_____	resist _____	_____
charge _____	_____	reproduc _____	_____

Your instructor will give you a spelling test.

VOCABULARY: GRAD/GRESS Roots

The root words GRAD and GRESS both come from the Latin word meaning "going" or "stepping." We generally use the GRAD- spelling when the root is at the beginning of the word, and -GRESS when the root is at the end of the word.

Gradual	going step by step, slowly
Gradient	the rate of change
Graduate	a person who completed a course of study
Aggressive	taking quick, sometimes violent action
Congress	to gather together
Digress	to go off in another direction
Progress	to go forward
Regress	to go backward
Transgress	to overstep or break a law

The familiar word GRADE also comes from this root. It literally means a step, stage, or level.

Grade	level or slope
Degrade	to lower in status or demote
Centigrade	divided into 100 degrees
Retrograde	going backward (adj.)

Exercise 14.9 Match the letter of the definition for each word.

_____ 1. digress   a. to overstep the law

_____ 2. congress   b. slope showing the rate of change

_____ 3. gradual   c. going backward

_____ 4. degrade   d. going off in another direction

_____ 5. gradient   e. moving ahead slowly

_____ 6. transgress   f. person who finished a program

_____ 7. progress   g. taking quick or violent action
_____ 8. retrograde   h. to demote
_____ 9. aggressive   i. to gather together
_____ 10. graduate   j. to go forward

Exercise 14.10 The following words are related to the words above. Write a definition (without using the dictionary) that relates the word to the root word.

Hint: The -ION suffix turns the word into a noun.

1. gradation _____
2. congregation _____
3. transgression _____
4. regression _____
5. graduation _____
6. degradation _____
7. progression _____
8. aggression _____

Hint: The -IVE suffix turns words into adjectives.

9. regressive _____
10. progressive _____

WORD WATCH: Stationary/Stationery Compliment/Complement

STATIONARY means not moving or fixed. Think of "stAnding still" to remind you that this word ends with ARY.

    Hold the probe stationary when reading the voltage.

STATIONERY means writing materials. Think of "lettERs" to remind you that this word ends in ERY.

    The letter was written on company stationery.

COMPLIMENT is a verb meaning expressing courtesy, praise, or flattery, or it is a noun meaning an expression of courtesy, praise, or flattery.

    I compliment you on your remarkable achievement. (*verb*)
    Please accept our compliments for your achievement. (*noun*)

The adjective form is COMPLIMENTARY.

    Include a complimentary closing on all letters. (*adjective*)

COMPLEMENT is usually a noun meaning something that fills or completes, or something added to complete a whole. Use the word "complete" to remind you of the middle E.

    The complement of a 45-degree angle is another 45-degree angle—together they total 90 degrees. (*noun*)

Business Letters

The adjective form is COMPLEMENTARY.

Complementary colors, when combined in the right intensities, will produce white light. (*adjective*)

Exercise 14.11 Use the correct word to complete each sentence.

Use STATIONERY OR STATIONARY.
1. The guard stood _____ while the crowd passed.
2. The _____ was white with black lettering.
3. The _____ machine collected dust.
4. The company's logo was printed on the _____.
5. How many envelopes were included with your _____.

Use COMPLIMENT or COMPLEMENT.
6. Some people do not know how to react to a _____.
7. Carol _____ed her subordinate's positive attitude.
8. A red tie will _____ a white shirt and dark suit.
9. A full _____ of personnel work the night shift.
10. His annual review was _____ary, and he was promoted.

CHAPTER FIFTEEN

REVIEW

I

READING: Ride the Tech Wave or Be Swamped by It

At the end of March, two Democratic U.S. senators—Sam Nunn of Georgia and Frank Lautenberg of New Jersey—introduced a bill that would create a bipartisan Information Age Commission.

The commission's job would be to determine what it will take to maximize the benefits computers and communications have on society; to maintain the U.S. lead in the world market for information; to equip individuals for an information society; and to encourage new technology.

In addition, the panel would study the impact of computers and communications on national defense, labor, and employment.

The bill calls for a $3 million budget and a two-year life span for the commission, which would have 23 members appointed by Senate and House leaders, and the president.

Of the president's 17 appointees, six would come from the executive branch of the federal government, including the secretaries of commerce, defense and education, or their representatives. The others would represent state and local government, and the private sector, including the information industry, labor, and academia.

The following is based on comments Lautenberg made in introducing the bill:

Our economy and our society have been transformed.

More than 60 percent of our nation's work force is employed in the creation, storage, processing or distribution of information, compared with just 17 percent in 1950. We were once an agrarian society. Then we became an industrial society. Now we are an information age society.

Leading this transformation are the communications and computer industries. Computing ser-

FRANK LAUTENBERG

Panel formed to help U.S. stay afloat in information age

"Ride The Tech Wave or Be Swamped by It," by U.S. Sen. Frank Lautenberg, *InformationWEEK,* May 6, 1985. Copyright © 1985 by CMP Publications, Inc., 600 Community Drive, Manhasset, NY 11030. Reprinted with permission from *InformationWEEK.*

vices and software industries are growing at a phenomenal 22 percent a year in the United States—and almost as fast in Japan and Europe.

The market in telecommunications products is projected to rise from $18 billion in 1983 to $41 billion 10 years later. Production of computers is rising at 10 percent a year.

We are a wired nation. Computer and telecommunications advances have affected the way we work, the way we play, and the way we learn. With a terminal and a data link, people can—and are—working from their homes, tapped into a firm's headquarters miles away. Children fortunate enough to have access to computers are learning in ways we never imagined.

Computer technology is revolutionizing the factory floor, assisting in the design and manufacture of products. From the broadest perspective, changes are occurring that will have a profound effect on our society and our place in the world economy.

We can ride the wave of this technological change, or we can be swamped by it.

The information age brings great benefits: increased productivity, enhanced communications, accelerated advances in science and technology. But it poses great challenges. There are challenges to our privacy, challenges to our industrial competitiveness, challenges to our ability to educate our youth, and challenges to our ability to maintain an active and self-fulfilled work force.

Congress has begun to address some of the critical issues presented by this transformation. Sen. Nunn has been a leader in addressing the complexities of computer crime. I have called for a national commitment to computer education, and to highlight the need, in a technology-based world economy, for an effective system of intellectual property rights to protect our nation's technological and economic edge. Our colleagues have begun to address issues of personal privacy.

Yet we lack a comprehensive review of the problems. We lack a comprehensive review of the choices we will face.

The Information Age Commission Act of 1985 would establish a commission to help us understand more clearly the complex issues and choices presented by the advent of the information age. The commission would contribute to the task of conducting such a review, provide a forum for the public discussion of these important issues, and synthesize work and thinking that has already been done. It would present the Congress and the nation with its analysis of the critical choices we face.

Technological change is proceeding at an accelerating rate. Policymakers, individuals, workers, companies, and society as a whole have less and less time to understand the impact of those changes, and to adjust.

The Information Age Commission is intended to contribute to our knowledge, our understanding, and our ability to make the changes wrought by the information age changes for the better.

Reading Comprehension Questions

1. What is the purpose of the Information Age Commission?

2. From what industries/areas would appointees to the commission be chosen?

3. Explain Senator Lautenberg's statement, "We are a wired nation."

4. What are some benefits of the information age?

5. What are some challenges of the information age?

6. What benefits have you experienced from the wave of technological change?

7. What is a danger that concerns you about the increased amount of technology in our society? Explain.

REVIEW: Chapters 13 and 14

Exercise 15.1 Proofread the letter below for spacing errors. Draw arrows to move text up or down a line or to move words to the right or left. Type the corrected letter.

Remember the following rules:

- Leave one space between words.
- Leave one space after most punctuation marks (except hyphens, quotes, parentheses).
- Leave two spaces between sentences.
- Skip lines between paragraphs.
- Skip lines between sections of letters.
- Indent a consistent number of spaces.

---

415 Old Orchard Drive

Redrock , AL 36866

September 5 , 1988

Radio-Electronics
    Reprint Department
500-B Bi-County Boulevard
Farmingdale , NY 11735
Dear Reprint Department:

   I would like to order   the booklet "All About Kits" advertised in the May 1986 issue of *Radio-Electronics.*

   I am enclosing acheck to $3.00 ($2.00 for the booklet and 1.00 for shipping and handling). Please mail   the booklet to the address given above . Thank you.

Sincerely,

Audrey Thompson   Audrey Thompson

Exercise 15.2 Type a letter to a local or national politician explaining your views on the challenges of the information age. Be sure to include all the parts of a business letter. Discuss one of the following:

- challenges to our privacy
- challenges to our industrial competitiveness
- challenges to our ability to educate our youth
- challenges to our ability to maintain an active and self-fulfilled work force.

Review

In your discussion, include what you consider to be the dangers presented by advanced technology and the controls you feel will eliminate these dangers. Proofread your letter carefully.

Exercise 15.3 Type a letter to a customer explaining his bill based on the following service repair order. Be sure to include all the parts of a business letter. Use the letterhead stationery for your final letter.

Apple Computer, Inc.

Service Repair Order — PLEASE COMPLETE ALL AREAS —

CUSTOMER INFORMATION
- NAME: Joe Brown
- ADDRESS: 3263 Trail Ct.
- CITY: Lilburn, GA
- STATE, ZIP: 30245
- TELEPHONE: 633-0000

DEALER INFORMATION
- Rick's Educational Services
- 3400 Oakcliff Rd. Suite 120
- Doraville, GA 30340
- Dealer No. 10027-6371

SRO NO.: G 61714
SPECIAL REFERENCE NUMBER: _____

TODAY'S DATE: 12/18/86
PRODUCT REQUIRING SERVICE: Apple II-e
SERIAL NUMBER: 70003721-0863
PURCHASE DATE: 12/1/86

OTHER EQUIPMENT (IF ANY): Disk Drive II, Monitor II

AppleCare AGREEMENT NUMBER: 973002
EXPIRATION DATE: 12/1/87

WARRANTY INFORMATION - Always Check Only One
- [ ] 1 - 30 DAYS
- [ ] 31 - 60 DAYS
- [ ] 61 - 90 DAYS
- [X] AppleCare
- [ ] SERVICE WARRANTY REPAIR
- [ ] SERVICE STOCK REPLACEMENT
- [ ] STORE DEMO REPAIR
- [ ] OUT OF WARRANTY
- [ ] OTHER: _____

TROUBLE REPORTED: Computer System will not load information from the disk drive. It will not "Boot-Up"

ESTIMATES — TIME: 1.5 hr. — COST: 52.50

DEALER COMMENTS: Replaced Stop-Button in disk drive. Set Drive Speed to specifications. Cleaned Disk Drive head Assembly.

QTY	PART NUMBER	DESCRIPTION	AMOUNT
1	1000-A24	Stop-Button	3.85
1		Labor @ 1.5 hr.	52.50

REPAIRED BY: X Jack Thompson

5353062600661714

- TOTAL PARTS: 3.85
- TAX: .15
- LABOR RATE: 1.5 HOURS x 35.00/hr RATE = 52.50
- TOTAL: 56.50

DEALER FILE

REORDER INFORMATION (408) 354-7792 - R. MILLER
FORM # 073-0101-B

 **Ricks Educational Services**

3420 Oakcliff Road, Suite 111, Atlanta, Georgia 30340
404-451-5181 or 424-7113

Exercise 15.4 Type or handwrite a letter of sympathy to a co-worker after his/her house was destroyed by a fire. Offer a specific type of assistance.

Exercise 15.5 Type a letter of recommendation for a student in your class who is applying for a job. Be sure to include the introduction, body, and conclusion in your letter.

Exercise 15.6 Type a memo to your supervisor reporting the status and projected completion date of a lab project. Be sure to include specific details and the four main parts of a business memo.

Exercise 15.7 Complete a troubleshooting report. Base your report on an actual experiment or assignment. Use only common abbreviations. Add information to make the report complete. Submit the following trouble log.

TROUBLE LOG

Customer			System Type		Serial No.	
	Date	Time	Problem Symptoms (Customer):			
Problem First Noted						
Called						
Arrived						
SVC Complete			☐ System Down	☐ Option Down	☐ Application Down	
Option			LARS No.		OK'd By	
Problem Diagnosis						

Exercise 15.8 Complete the following crossword puzzle using words from the Vocabulary and Word Watch sections in Chapters 13 and 14.

ACROSS

5. to replace (13)
7. former time (13)
8. to advance (13)
12. something that completes (14)
13. going backward (14)
15. to go off in another direction (14)
18. a level or slope (14)
19. to go beyond or surpass (13)
20. to separate (13)
21. praise or flattery (14)

Review

DOWN

1. to go backward (14)
2. to go back (13)
3. handed or moved (13)
4. not moving (14)
6. to go forward (14)
8. to go in front of or before (13)
9. an amount more than is needed (13)
10. to demote (14)
11. writing materials (14)
14. slow going (14)
16. rate of change (14)
17. to turn out as planned (13)

Choose from the following words:

complement	passed	secede
compliment	past	stationary
degrade	precede	stationery
digress	proceed	succeed
exceed	progress	supersede
excess	recede	
grade	regress	
gradient	retrograde	
gradual		

Exercise 15.9 Study the following spelling words from Chapters 13 and 14. Your instructor will give you a spelling test.

available	audible	supersede
dependable	tangible	cede
excitable	invisible	secede
usable	feasible	intercede
appreciable	collectible	antecede
sociable	reversible	precede
demonstrable	suggestible	concede
practicable	flexible	recede
serviceable	collapsible	exceed
chargeable	resistible	proceed
noticeable	reproducible	succeed

PART
V

GRAMMAR UNITS

1 *Subjects and Verbs*
2 *Fragments*
3 *Compound Sentences*
4 *Complex Sentences*
5 *Subject/Verb Agreement*
6 *Prepositional Phrases*
7 *Pronouns*

8 *Pronoun Reference*
9 *Modifiers*
10 *Parallelism*
11 *Avoiding Shifts*
12 *Avoiding Sexism*
13 *Transition Words*

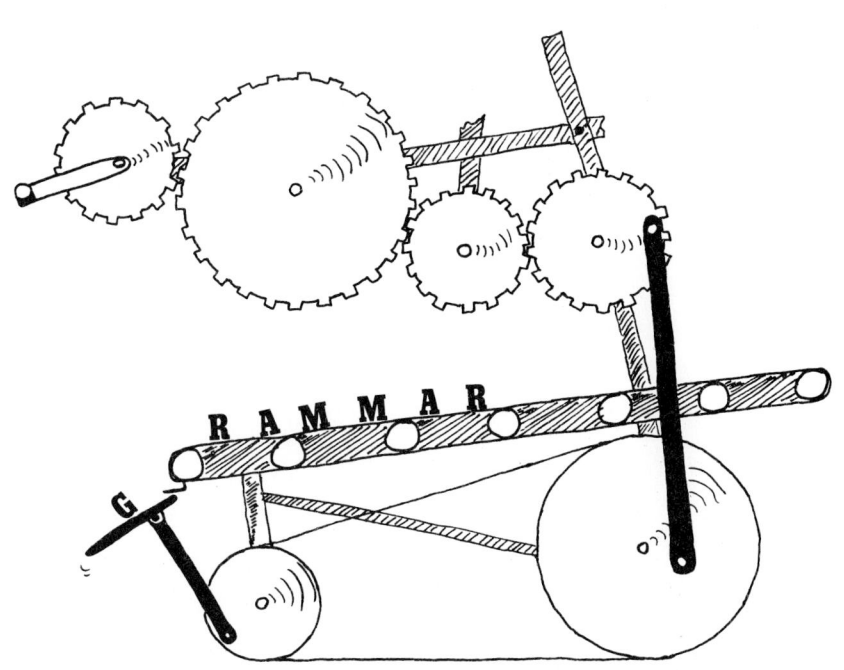

GRAMMAR UNIT ONE

SUBJECTS AND VERBS

1

The most fundamental elements of a sentence are the **subject** and **verb.** Without both, a written idea is incomplete and cannot be called a sentence. A subject can be either a noun or a pronoun, and a complete verb may consist of up to three verbs, either action or linking. We begin this review with a careful look at nouns.

NOUNS

Most people can remember the informal definition of a **noun:** a person, place, thing, or idea. The first three types (persons, places, and things) are the easy nouns to spot. They can be singular or plural. The plural is usually formed by adding S or ES.

The final type of noun is the idea. In fact, the word "idea" is a noun. Other examples of idea nouns are found in the following sentence.

>   *Example:* Your *education* and technical *knowledge* will be useful in an electronics *career.*

You can see that the italicized nouns in the example above are ideas, not things that we can see or touch. One test is to say the word with A, AN, or THE in front of it.

>   an education
>   the knowledge
>   a career

If doing so makes sense, it is a noun. If not, it is probably some other kind of word.

You will probably remember that **pronouns** are similar to nouns. They take the place of nouns. In these exercises, we do not consider pronouns as nouns. We deal with pronouns in a later chapter.

Exercise 1 Underline the nouns in the following sentences. The number of nouns in each sentence is given in parentheses.

1. Resistance is the opposition to current. (3)
2. This quality is used in electrical circuits to limit or control the amount of current that flows. (4)
3. When current flows through resistance, heat is produced. (3)
4. In certain applications, the main purpose of resistance is to produce heat. (4)
5. In many other applications, the heat produced represents an unwanted loss of energy. (4)
6. Resistance is measured in units of ohms. (3)
7. One ohm is the resistance when one ampere flows with one volt applied. (4)
8. An electrical component having the property of resistance is called a resistor. (4)
9. There are many types of resistors in common use, but generally they are placed in two main categories. (4)
10. The two main categories are fixed resistors and variable resistors. (3)

Exercise 2 Write your own nouns in the following categories:

People	*Places*	*Things*	*Ideas*
Name people famous for electronics principles.	Name cities where you would enjoy living.	Name equipment you use in the lab.	Name qualities of electronic devices (noun forms end in "ance").
_____	_____	_____	_____
_____	_____	_____	_____
_____	_____	_____	_____
_____	_____	_____	_____

SUBJECTS

Subjects are almost always nouns or pronouns (words that take the place of nouns). They are usually found near the beginning of a sentence. They tell who or what the sentence is about. In the following example below, the subject is italicized.

Electronics is the study of electron movement.

The sentence is about "electronics," and we will call that word the "subject" of the sentence. There are other nouns in the sentence, but only one subject.

Some sentences have more than one subject, and we call these "compound subjects." The compound subjects are italicized in the following example.

Television, radio, and *radar* are some applications of electronics that we take for granted.

Subjects are usually written at the beginning of a sentence, although they can appear at other places in the sentence, too. The delayed subject is italicized in the following example.

To understand electronic devices, a *person* must know something about electronic current.

The sentence above is not about devices (the first noun in the sentence), but about a person.

Exercise 3
Draw a line under the subject/s of each sentence.

1. Coulomb measured the amount of these forces.
2. He proposed the law now known as Coulomb's law.
3. The force is proportional to the product of the charges.
4. It is inversely proportional to the square of the distance.
5. We cannot explain this subject or begin an examination.
6. The arrangement, electric charge, mass, motion, and energy must be introduced.
7. Called a "building block" of matter, the atom is made up of many smaller particles.
8. We can refer to the model of Bohr and Rutherford.
9. It represents an atom as a loose structure of electrons surrounding a nucleus.
10. The nucleus is composed of particles of a positive charge.

VERBS

A **verb** is usually defined as a word showing action—what *is, was,* or *will be happening*. All the following words are action verbs, things you can do.

      LEARN   CONNECT   FOLLOW   WIRE   REFER   DISPLAY

Verbs have the special job of indicating the time of the action by **verb tense**. There are three main tenses: *past, present,* and *future*. The tense is changed when the writer changes the time:

Years ago, technicians *recorded* on notepads.
Today, technicians *record* on service report forms.
In the future, technicians *will record* on computers.

HELPING VERBS AND LINKING VERBS

Another group of verbs, called **helping verbs,** are commonly used in combinations with each other or action verbs. Some of them, called **linking verbs,** can be used alone, and others (such as *be* or *been*) cannot be used alone.

AM	IS	ARE	WAS	WERE	BE	BEEN
HAVE	HAS	HAD	DO	DOES	DID	WILL
SHALL	CAN	MUST	WOULD	COULD	SHOULD	MIGHT

Whenever you see these verbs, they are either the verb or part of the verb in a sentence. It is the only job these words do, unlike most other English words. Sometimes they are followed by action words, and other times they are used alone.

In an MBM, data *are stored* in the form of magnetic bubbles.
    (helping verb + action verb)
The bubbles *are* cylindrical magnetic domains.
    (linking verb used alone)

PARTICIPLES

There are two other common types of verbs, both called participles. One is the **present participle,** or -ING form, and the other is the **past participle,** or -ED form. Both of these verb forms require a helping verb. They add a *continuous* time frame, an extended present or past time period, to sentences.

> -ING form: Technicians *are using* integrated circuits.
>   *(extended present use)*
> Past participle: Technicians *have used* integrated circuits.
>   *(extended past use)*

Regular verbs have predictable patterns.
To change from *present* to *past* tense, add -ED to the end of regular verbs.

> Today we *tune* the car. *(present)*
> Yesterday, we *tuned* the car. *(past)*

To change to the *future* tense, add "will" in front of the present-tense verb.

> We *will tune* the car tomorrow. *(future)*

To use the *past-participle* form, add a helping verb to the -ED (past-tense) verb.

> We *have tuned* the car several times. *(past participle)*

To use the *ING* form, use a helping verb and add -ING to the present-tense verb.

> We *are tuning* the car. *(ING form)*

Note   Remember to write the final "ED" on past- or participle-tense verbs even if you do not pronounce them when you speak. Leaving them off in writing makes you sound uneducated and careless.

> *Wrong:*   I am suppose to be in class.
> *Correct:*   I am supposed to be in class.
>
> *Wrong:*   We use to have lab on Wednesday.
> *Correct:*   We used to have lab on Wednesday.

Irregular verbs have unusual forms in the past- and past-participle tenses and have to be remembered. If you have ever heard a young child say, "We runned down the street," you recognized that the child overgeneralized the ED rule, and will eventually learn about irregular verbs. Common irregular verbs are listed in Appendix 5. The past tense is formed in some other way than by adding -ED. Notice that the last column is the participle form, which is used following a helping verb. "Bring" is an example of an irregular verb.

> Today, I *bring* suggestions. *(present)*
> Yesterday, I *brought* nothing. *(past)*
> I *will bring* all the connectors. *(future)*
> I *have brought* everything for the circuit. *(past participle)*
> I *am bringing* the power supply. *(ING form)*

Subjects and Verbs 205

---------- **NOTE** ----------

The jargon word "troubleshoot" has NO past tense. When referring to the past, writers have to reword sentences to use the present or ING form of this word.

Examples

As a result of troubleshooting, we discovered the problem.
After troubleshooting, we discovered the problem.

Exercise 4 Underline the complete verb (using *two* lines) in each sentence. Underline the subject using *one* line.

Example: The technician studied the resistor.

1. The ohmic value is color-coded on the resistor.
2. The resistor is read from left to right.
3. The ohmic value may be very large.
4. Resistances might be as high as millions of ohms.
5. Metric symbols are used to represent thousands and millions.
6. The value can be written using symbols.
7. Only resistance can dissipate electrical power.
8. The friction causes heat to develop in the circuit.
9. Resistors must be able to dissipate this heat.
10. Heat is measured in units called watt-seconds.

INFERRED YOU
In the following sentence, the subject is not written as a word.

    Connect the transistor.

The action, or verb, is "connect." The person who will do the "connecting" is the reader. We call this type of subject the **inferred you.** The verb is always in the present tense. We use this type of sentence only for commands or instructions.

Exercise 5 Underline the subjects once and the verbs twice in the following sentences. If the subject is the inferred "you," draw a star (*) above the present-tense verb.

1. Many careers in electronics are possible.
2. Watch for the diverse opportunities available for electronics technicians.
3. Some companies maintain electronics laboratories for research and development.
4. Read the indexes for information on specific companies.
5. Don't be led to believe that technicians only do troubleshooting.

Exercise 6 Supply the correct form of the given verb to fit the sentence. Refer to Appendix 5 for irregular verb forms.

1. (see)   I have _____ a great change in my soldering skills.

Subjects and Verbs

2. *(swear)* When I burned my arm on the soldering iron, I _____ I'd never solder again.
3. *(write)* I have not _____ anything longer than a three-page report.
4. *(cut)* I _____ part of the report after I read it.
5. *(take)* It has _____ me several hours to read the operating manual.
6. *(break)* I _____ my concentration when a friend came.
7. *(do)* The report was _____ before midnight.
8. *(give)* I have _____ up my habit of putting things off.
9. *(prove)* The argument could not be _____.
10. *(throw)* After the report was returned, I _____ my backup copy away.

Exercise 7 In the following article, underline subjects once and verbs twice. Then, rewrite the article by changing the present-tense verbs to past tense.

ELECTRONIC HEARSAY

The U.S. Justice Department is itching to enlist electronic ingenuity in its fight against white-collar crime. It wants to establish computer files that allow police to exchange information on persons they believe are involved in furtive enterprises such as fraud or embezzlement. Although the computerized system is possible, this project comes with red flags firmly affixed. It does not restrict information collected to matters of public record. Instead, unverified data can be contributed and perused by police officials with access to the files.

Exercise 8 Use each present tense verb in a sentence. Then rewrite the sentence, changing the verb to past, future, past participle, and ING forms.

Example: Come
  a. *(present)* We come to school.
  b. *(past)* We came to school.
  c. *(future)* We will come to school.
  d. *(past participle)* We had come to school.
  e. *(ING form)* We are coming to school.

1. Break
   a. (present) _____
   b. (past) _____
   c. (future) _____
   d. (participle) _____
   e. (ING form) _____
2. Go
   a. (present) _____
   b. (past) _____
   c. (future) _____
   d. (participle) _____
   e. (ING form) _____

3. Write
   a. (present) _____
   b. (past) _____
   c. (future) _____
   d. (participle) _____
   e. (ING form) _____
4. Do
   a. (present) _____
   b. (past) _____
   c. (future) _____
   d. (participle) _____
   e. (ING form) _____
5. Bring
   a. (present) _____
   b. (past) _____
   c. (future) _____
   d. (participle) _____
   e. (ING form) _____

SENTENCE COMBINING

Sentences can be as short as one word. In the sentence "Stop," there is only a verb, and the subject is the "understood you," the person or people being addressed. Most sentences are longer. Sentences can be much longer as a result of combining ideas. The following passage is an example of a long sentence.

> In 1785, Coulomb carried out a series of experiments on small charged bodies during which he observed that two spheres, each with an electric charge, repel each other when the charges are of the same sign.

This passage could be reduced into the following smaller sentences, more like the writing style in grade-school science books.

> In 1785, Coulomb carried out a series of experiments.
> The experiments were on small charged bodies.
> He observed that spheres have an electric charge.
> Sometimes two spheres repelled each other.
> This happened when the charges of the two spheres were of the same sign.

The ideas in the short sentences deliver the same message as the one longer sentence. Notice that the group of short sentences has more words, and some ideas have been repeated. Although simplicity is often effective, technical writers use as few words as possible and avoid unnecessary repetitions.

One method of practicing efficient, clear wording is to combine short sentences into complex, powerful sentences. The purpose is to reduce the number of words. You will be given a group of short, related sentences, and you will reword them into one sentence. Remember that there is no one correct way of combining. In fact, you will perhaps see several possible ways of combining each group. The best combination is the one that delivers the message clearly and correctly. Consider the following short sentences.

208   *Subjects and Verbs*

Example: a. A computer system consists of areas.
b. There are three main areas.

Two possible combinations are:

A computer system consists of three main areas.
There are three main areas in a computer system.

Both of these sentences are grammatically correct; however, the first sentence is written in the active style, which is preferred in technical writing. It introduces the subject (computer system) first, has an active verb, and has one fewer word. Avoid using "There are . . . " at the beginning of sentences.

When combining sentences, *use the first sentence as the core.* You may add or eliminate some words, but do not change or add to the intended meaning. You will frequently be reordering and reducing the number of words. For instance, "the man is young" could easily be reordered "the young man . . . " to allow more ideas to be added.

———————— **CAUTION** ————————

It is possible to link each sentence with "and," but this would not reduce the number of words.

Exercise 9   Combine each group of sentences into one complex sentence. Do *not* link them only by using AND, and do *not* change or add to the meaning. You may add or change words if necessary, and you may reorder the words and ideas.

THE COMPUTER SYSTEM

1. a. A computer system consists of areas.
   b. There are three main areas.
   c. One area is input.
   d. One area is processing.
   e. One area is output.
   Combination: _____

2. a. There is a fourth area.
   b. This area is storage.
   c. It backs up the computer system.
   Combination: _____

3. a. The system is made up of hardware.
   b. The hardware is machines.
   c. The system is made up of software.
   d. The software is programs.
   Combination: _____

4. a. The programs turn data into information.
   b. The data are unprocessed.
   c. The information is usable.
   Combination: _____

5. a. Data may be collected.
   b. Data may be processed.
   c. It may happen immediately.
   d. It may happen later in groups.
   Combination: _____

Check your sentences with your instructor.

GRAMMAR UNIT TWO

FRAGMENTS

A group of words that has a subject and verb is called a **clause.** A clause that also has a complete idea is called an **independent clause,** or a **sentence.**

A sentence must have a complete subject and a complete verb, and it must express a complete idea. A group of words without a complete subject, verb, or idea is called a **fragment**.

We can speak in fragments and still communicate effectively. This is due mainly to other cues besides our words, such as eye and hand movements and voice inflection. We can observe and question our listeners, and we have the luxury of repeating and rewording our message until we are satisfied it has been understood correctly.

In writing, however, we cannot observe our readers. In professional writing, we must make every attempt to follow standard English rules to ensure effective, efficient communication. Using complete sentences is an important rule.

A fragment may be caused by many types of errors: a missing subject, a missing or incomplete verb, or an incomplete thought. In this unit you will learn to identify and correct three types: missing subject, missing verb, and phrase fragments (see Grammar Unit Four for a review of incomplete idea fragments).

MISSING SUBJECT FRAGMENTS
The subject of the sentence may appear in various places in a sentence.

> Personal *robots* hold promise of creating greater excitement than the computer.
>
> So where are *all* these robots?
>
> Once found only in industry and the laboratory, over the past decade the *computer* has become a fixture in the majority of households.

The subject tells what the sentence is about. The subject may be a noun (person, place, thing, or idea) or a pronoun (he, she, it, they). Pronouns are used in paragraphs only after the noun has been stated. A verb ending in "-ing" can function as a noun and the subject of a sentence.

> *Mowing* the lawn, *washing* clothes, and *taking* out the garbage are some of the obvious uses.

Note   The inferred "you" subject is used only with present tense verbs.

> *Wrong:*   Took a chance.
> *Correct:*   (You) Take a chance.

Exercise 1   Supply a subject for each sentence.

1. _____ is the study of electron flow.
2. _____ taught my first electronics class.
3. _____ is a company at which I intend to apply for a job.
4. _____ likes to study with a partner.
5. _____ is the course in which I learned the most.

Exercise 2   Underline the subject of each sentence. If the subject is missing, rewrite the sentence to make it complete.

1. Can take any form.
2. By the mid-1980s, hundreds of thousands of people owned personal computers.
3. After the novelty wore off, wound up back in the closet.
4. The reality was that only a small number of people had a real need.
5. For everyone else, was a device without an application.
6. With personal robots is very different.
7. Notice the many applications.
8. Aren't capable of performing more than a few instructions.
9. Reading this survey of the personal-robot field has been helpful.
10. Found out when and what form.

MISSING VERB FRAGMENTS

The verb in a sentence states an action or condition (state of being) about the subject.

> Heath Company *introduced* the Hero Jr in mid-1984. *(action)*
> The Hero Jr *was* basically a toy. *(state of being)*

Including the helping verb with participles not only sounds educated, but ensures that the correct message is being sent. In the following sentence, notice that the absence of a helping verb changes the meaning.

> The Hero Jr *introduced* by Heath Company.

The Hero Jr did not introduce anything, but it "was introduced." Often in speaking, we slide over helping verbs (or the final ED or S), which starts a careless habit of

212 *Fragments*

leaving them off in writing. A present (ING) or past-participle verb without a helping verb is a fragment. The helping verb can be added as a word or contraction.

Exercise 3 Underline the complete verb. If the verb or part of the verb is missing, rewrite the sentence and make it complete.

1. Of the companies currently producing robots, Heath done well.
2. The personal robots industry unstable.
3. New manufacturers appearing and disappearing all the time.
4. All the mechanical parts were designed to have multiple functions.
5. All the products available for general sales.
6. In November 1985, the division of Ideal responsible for Maxx Steele.
7. We followed the same steps.
8. A time line of the history of personal robots versus that of the PC.
9. Firms beginning to enter the market.
10. A Texas-based marketing-research firm seen the investigation.

LINKING FRAGMENTS TO SENTENCES

Another way of correcting missing-subject or missing-verb fragments is to combine the fragment with a related sentence. A sentence may have two subjects or two verbs, but there must be a logical connection in which the fragment adds to the meaning of the sentence.

Wrong: The robots of today. And workers perform side by side.
  (missing-verb fragment) (sentence)
Correct: The robots of today and workers perform side by side.
  (sentence with two subjects)

Wrong: Robots are available. And used for many purposes.
  (sentence) (missing-subject fragment)
Correct: Robots are available and are used for many purposes.
  (sentence with two verbs)

──────────── **WARNING** ────────────

Do not add commas between a pair of subjects (*compound subjects*) or a pair of verbs (*compound verbs*).

Exercise 4 If the following groups contain one or more fragments, combine the fragment(s) to make a complete sentence and write the combination. Change the punctuation to make the combination correct. If there is no fragment, write CORRECT.

1. The installed base of robots in the United States. Is seen in many applications and industries. _____
   _____

2. The robot is a powerful tool in the manufacturing area. And requires specialized environments. _____
   _____

3. Robots are unforgiving. They reveal inefficient, ill-conceived, and outmoded processes. _____
   _____

Fragments **213**

4. Robot manufacturers are aggressively looking to improve their products. And broaden their market. _____

5. Enhancements such as vision, sensors, and communications systems. Are being applied almost routinely. _____

6. Robot manufacturers are not sitting on their haunches. They want to assure the robot's position on the market. _____

7. Too often the robot has been looked on as a human replacement. And subjected to environments designed for humans. _____

8. The conditions of the workstation. And inherently human skills contribute to the human operator's low productivity. _____

9. The robot, in a very real sense. Forces the manufacturer to look at how process operations are organized. _____

10. Industry leaders hope for sustained growth of robot usage. And are committed to increased development. _____

PHRASE FRAGMENTS

A third type of fragment, the **phrase fragment,** suffers from the lack of a subject and a verb. It may be made up of one or more prepositional phrases. Remember that subjects are never inside prepositional phrases. These fragments can be corrected by combining the fragment with its parent sentence.

Wrong: Robot sales have fallen short. Of earlier forecasts.
 (sentence) *(phrase)*
Correct: Robot sales have fallen short of earlier forecasts.
 (sentence with phrase)

Exercise 5 Write the combination necessary to correct any phrase fragments. If there is no fragment, write CORRECT.

1. To understand the situation properly. One must keep the current state of robotics in perspective. _____

2. First, the technology continues to advance. To the benefit of the user both in terms of reliability and capabilities. _____

3. A sure sign of industry maturity is the serious work being done. Several committees protect the standards for robot design. _____

4. Second, the robot becomes an integrator. Of a variety of advanced manufacturing technologies. _____

5. Third, the growth of the robotics industry is tied to an attempt. To improve and modernize manufacturing processes. _____

6. We are aware that the manufacturing process will be renewed. But we lack the commitment to get the job done. _____

7. The robot is a tool. Much like CAD/CAM, machine vision, and a host of other technologies. _____

8. They are interrelated. Through marketing and development, their benefits are known. _____

9. They will proliferate only at the rate we choose. To integrate and apply them. _____

10. By learning more about robotics. This technology can grow and prosper. _____

Exercise 6 There are six fragment errors in the following paragraph. Find the fragments. Then edit the paragraph to correct the fragments by combining them with sentences and adding or removing periods and capital letters.

The first International Personal Robot Congress and Exhibition held in 1984 in Albuquerque, New Mexico. Isaac Asimov, author of hundreds of books. Opened the event. Spoke of robots as they have appeared in his science fiction and as he thinks they should be in reality. Asimov laid down the "three rules of robotics" which Asimov claims still stand: (1) A robot not injure a human being, or through inaction allow a human being to come to injury. (2) A robot must obey orders given to it. By a human being, except when those orders would violate the first law. (3) A robot must protect its own existence. Except when that would violate the first or second law.

GRAMMAR UNIT THREE

COMPOUND SENTENCES

A group of words with a subject and verb is called a **clause.** A clause with a complete idea is called an **independent clause,** or a **sentence.** A sentence that has just one independent clause is called a **simple sentence.** The sentences below are simple sentences.

> A picture is worth a thousand words.
> In the business world, words rule.

clause = group of words with a subject and verb
independent clause = *sentence*
one independent clause = *simple sentence*

If we wrote only simple sentences, our writing would appear choppy and dull. To make our writing smoother and more interesting, we vary our sentence structure. One way of doing this is joining two sentences together to make compound sentences.

A **compound sentence** has more than one independent clause joined with a *connecting,* or *coordinating,* word.

> A picture may be worth a thousand words, but in the business world, words rule.

two or more independent clauses = *compound sentence*

Coordinating words show that the two complete sentences are of equal importance, yet they are joined because the author wants to relate them.

216 *Compound Sentences*

Exercise 1 Read each compound sentence. Underline the two independent clauses in each sentence, then circle the coordinator. Notice that each clause has a complete subject and verb.

1. There are two basic types of graphics, and they are analytical and presentation.
2. No clear line of demarcation exists, but there are certain recognizable differences.
3. A spreadsheet can put numbers into many rows and columns, so it can give a large amount of information.
4. Getting images into the computer can bring added expense, for graphic input devices like a mouse or touchpad cost extra.
5. Most people are not content to gather a crowd around a screen to view graphics, so they find a way to print the graphics.
6. The hard copy may be in the form of a paper print, or it may be on acetate film for an overhead projector.
7. One way to store information in the computer is by bit mapping, and it is a useful way to store and use graphic data.
8. The other way is to keep graphics in a computer by means of mathematical relationships, so the computer remembers to store a line at a certain angle leading from one point to another.
9. Even photographs are composed of very tiny dots, yet, to the eye, photographic images appear to be solid.
10. In color printing, each color is printed separately, and, as a result, it is very slow.

The common coordinators, words used to join two sentences, are the following:

AND   BUT   FOR   NOR   OR   YET   SO

Do not confuse compound sentences with simple sentences containing compound subjects or verbs. Read the following sentences. Mark the subject(s) and verb(s).

> The degree and accuracy of control ranges from fair to excellent.
> *(simple sentence with compound subject)*
>
> Our brain's left side processes and analyzes logical information, data, and concepts.
> *(simple sentence with compound verb)*
>
> Artistic information and ideas are analyzed and processed in the right hemisphere of the brain.
> *(simple sentence with compound subject and verb)*

In the following exercise, practice recognizing the difference between compound sentences and simple sentences with compound subjects and verbs.

Exercise 2 Read each sentence. If it contains two independent clauses, a compound sentence, write C in the blank. Circle the comma and coordinator. If it is a simple sentence, write S in the blank.

Clue: Simple sentences may have compound subjects or verbs.
Compound sentences have two complete sentences.

_____ 1. Analytical graphics software allows you to note trends or relationships, and presentation graphics show concepts.

_____ 2. Businesses are recognizing this and are using more graphic representations in communications.

_____ 3. Printers, plotters, and graphics cameras generate sharp, colorful visual displays to aid you in presenting difficult-to-grasp information.

_____ 4. There is no doubt that graphics has arrived, as more than a million graphics software programs were purchased last year.

_____ 5. The ability to add text in many sizes and type styles, design special characters, and show a series of charts and graphs in a slide-show style are just a few of the features.

_____ 6. Programs can take data from existing spreadsheet or database files, or they can enhance graphs to improve their appearance.

_____ 7. A large percentage of business presentations is in the form of word or text charts, so programs that generate text charts in a variety of sizes and styles is a boon.

_____ 8. Another printer that is becoming popular, especially for color graphics, is the ink-jet printer.

_____ 9. Early ink-jet printers were not easy to use, nor were they easy to clean.

_____ 10. Creating high-quality, professional graphics is neither cheap nor easy, but the added impact makes the process worthwhile.

PUNCTUATING COMPOUND SENTENCES

An incorrectly punctuated compound sentence is similar to a cold-solder joint. The result is a poorly fused connection. To make sentences "fuse," we use punctuation as "solder." You also know that different types of solder are used depending on the characteristics of the circuit. Similarly, when fusing compound sentences, we have a choice of punctuation depending on the characteristics of the sentence. These choices are (1) a comma and coordinator, or (2) a semicolon.

Commas and Coordinators To complete a fusion, we can put a **comma and a coordinator** between the two independent clauses.

> A picture may be worth a thousand words, but in the business world, words rule.
>
> Our left hemisphere processes logical information, and our right hemisphere processes artistic information.

_____ **WARNING** _____

Remember that connecting compound sentences with only a comma or only a coordinator improperly fuses the two sentences.

Comma splice error: Businesses are recognizing this, they are using more graphic representations.

Run-on error: Businesses are recognizing this and they are using more graphic representations.

Correct: Businesses are recognizing this, and they are using more graphic representations.

Exercise 3 Fuse the two sentences into a compound sentence by adding a comma and coordinator. Be careful to use a logical coordinator. Remember to smooth the fused sentences when necessary, as in the example.

1. In any circuit, there is always the possibility of overcurrent.
   Overcurrents occur for many reasons.

   Example: In any circuit, there is always the possibility of overcurrent, and it occurs for many reasons.
   (Notice that the pronoun "it" is used in the second clause to avoid repeating "overcurrent." Be sure that the pronoun agrees in number with the noun.)

2. An overcurrent means that the current is too large.
   The connecting wires cannot be handled safely.
3. Overcurrent can produce a dead short circuit.
   Overcurrent can produce a partial short circuit.
4. Overcurrents may be produced if too many loads are switched ON at the same time.
   Most overcurrents occur because of some malfunction in the circuit.
5. Two supply wires may appear to be well separated.
   They may actually come very close to each other.
6. In partial short circuits, resistance does not drop to zero.
   Supply-line currents are not necessarily large enough to trip the circuit-protective device.
7. Partial short circuits can be caused by worn insulation.
   If moisture collects in the cracks, the insulation's resistance is greatly reduced.
8. On a humid day, cracked wire insulation may develop resistance low enough to trip the protective device.
   On a dry day it may not.
9. Particulate residue from the surrounding air may build up across the insulating barrier between wire terminals.
   A partial short circuit will result.
10. A protective device can open to disconnect an ungrounded wire from the source.
    This will stop the overcurrent.

Semicolons in Compound Sentences Another method of fusing compound sentences is using a semicolon (;) between the two clauses.

> Flatbed plotters hold the sheet of paper in place on a flat surface; the pens draw their lines moving in each direction to form the image.

Notice that the second clause *does not start with a capital letter and does not use a coordinator.* The second clause is considered a continuation of the first clause. Semicolons should not be overused or their effectiveness will be diminished. If semicolons are used carefully and appropriately, they provide an abrupt, striking coordination; overused, they become weak and showy.

Remember that semicolons can be used if the following conditions are met.

1. The two clauses are not coordinated by AND, BUT, OR, NOR, FOR, SO, or YET.

Wrong: A fuse is a protective device; and it consists of a metal link inside a casing.

2. The ideas of the two clauses are closely related.

Wrong: A fuse is a protective device; ungrounded-frame wiring systems are unsafe. *(unrelated ideas)*

The second clause may begin with connective words or phrases such as THEREFORE, HOWEVER, IN FACT, ON THE OTHER HAND, or other transitional phrases. However, these words or phrases may also be used to begin a new sentence. Sentence length is usually the determiner. If both clauses are long, use a period.

The two supply wires may appear to be well separated; therefore, the short circuit is difficult to locate and eliminate.

If an accidental short occurs between the hot supply wire and the frame, there is no path by which overcurrent can flow back to the source. Therefore, the circuit breaker does not open, and the top supply wire remains connected to the source.

Exercise 4 Rewrite the compound sentences in Exercise 3. Fuse them with semicolons.

1. *Example:* In any circuit, there is always the possibility of overcurrent; it occurs for many reasons.

2. _____

3. _____

4. _____

5. _____

6. _____

7. _____

8. _____

9. _____

10. _____

Check your punctuation with your instructor.

GRAMMAR UNIT FOUR

COMPLEX SENTENCES

A group of words with a subject and verb is called a **clause.** A clause with a complete idea is called an **independent clause** or a **sentence.** A sentence with two or more independent clauses is called a **compound sentence.**

    clause = group of words with a subject and verb
    independent clause = *sentence*
    two independent clauses = *compound sentence*

In this unit, you will practice combining clauses that are not both independent. One clause will be independent, now called the **main clause.** The other will be dependent, now called a **subordinate clause** because one idea will be *subordinate,* or *dependent,* on the main idea. The prefix "sub" means beneath or low. A subordinate clause will be less important than the main clause.

                  sub = low
         subordinate = lower in importance
  subordinate clause = incomplete idea fragment

  independent + dependent clauses = *complex sentence*
  main clause + subordinate clause = *complex sentence*

We call sentences with unequal clauses **complex sentences.** One purpose for using subordination is to add interest and variety to sentence structure. Another purpose is to add unity to sentences by explaining the relationship of one clause to another.

The words we use to signal subordinate ideas are called **subordinate conjunctions,** but they can be more easily remembered as **signal words,** since they give direction to the meaning of the sentence. Read the following sentence and notice how the signal word "because" relates the two clauses.

    The Solar Max project was declared a success *because* the mission objectives established before the launch were met.

If you take this complex sentence apart, you will find two clauses:

> The Solar Max project was declared a success
> *(main clause)*
> because the mission objectives established before the launch were met.
> *(subordinate clause)*

The signal word "because" makes the second clause depend on the first. The main clause of the sentence is independent and could correctly be written as a simple sentence. A subordinate clause by itself is a fragment because it does not express a complete idea. It needs an independent clause to complete it.

A subordinate fragment can also be corrected simply by removing the signal word.

> *Because* the mission objectives established before the launch were met.
> *(an incomplete fragment)*
> The mission objectives established before the launch were met.
> *(a complete sentence)*

Remember that every complex sentence needs at least one independent clause. A subordinate clause written alone is a fragment.

When we use subordination to show a relationship, that relationship is determined by the signal word. The subordinate clause can introduce or follow the main clause.

> The Solar Max project was declared a success *because* the mission objectives established before the launch were met.
> *Because* the mission objectives established before the launch were met, the Solar Max project was declared a success.

Writers usually alternate between the two methods throughout a report to add variety to their sentences. As a rule of thumb, however, remember that long, complicated technical sentences are usually more understandable if they begin with the main clause.

The most common signal words are listed below. Notice that each one is used to show a relationship.

Subordinate Conjunctions (Signal Words)

WHY	WHEN	HOW	WHERE
because	after	although	where
if	as	as if	wherever
since	before	as though	
so that	once	how	
	since	unless	
	until	though	
	when	even though	
	whenever		
	while		

SPECIAL SIGNAL WORDS			
that	which	who	
	whichever	whoever	

Complex Sentences

There are many types of complex sentences. Some have more than one main clause or more than one subordinate clause. Often, these sentences can be rearranged several different, correct ways. The writer must decide which way is most effective.

Exercise 1 Underline the main clause in each complex sentence. Circle the subordinate conjunction. Draw a wavy line under the subordinate clause.

1. When the space shuttle program was first conceived, one of its primary benefits was intended to be the retrieval or repair of satellites.
2. The Solar Max was chosen because it was specially designed.
3. If a regular socket wrench were used on these bolts, the astronaut himself would turn.
4. The ground controllers placed the satellite in a slow, stable spin after it failed.
5. Objects in space, although they are weightless, are not massless.
6. As any service technician knows, things rarely go as planned.
7. Grabbing a spinning satellite is no simple matter, as you might imagine.
8. The first task, then, was to stop that spin so that the shuttle's manipulator arm could pick up the satellite safely.
9. Once the bolts were removed, the service tool is used to hold the module.
10. The faulty module is stowed in a temporary location so that it can be returned to earth and examined to find the cause of the failure.

Exercise 2 Complete the sentences by adding your own dependent clause beginning with the signal word. Be sure that your added clause has a complete subject and verb. Make each sentence different and true.

1. Service technicians have interesting jobs because _____
   _____.

2. Service technicians have interesting jobs after _____
   _____.

3. Service technicians have interesting jobs unless _____
   _____.

4. Service technicians have interesting jobs although _____
   _____.

5. Service technicians have interesting jobs when _____
   _____.

COMMAS IN COMPLEX SENTENCES

One of the clauses in a complex sentence is the main (independent) clause. Recognizing this main clause is the trick to knowing when and where to place commas in complex sentences.

The rule of thumb for using a comma is as follows:

_____ **RULE OF THUMB** _____

If the subordinate clause introduces the main clause, put a comma after the subordinate clause.

Do not put a comma after a main clause when it is followed by a subordinate clause.

Notice that the rules as stated above follow the rules. In "picture" form below, the signal word represents a subordinate conjunction, and the lines represent clauses. The signal word makes the clause that follows it dependent.

```
SIGNAL __DC__ , __IC__ .
__IC__        SIGNAL __DC__ .
```

Exercise 3 Underline the main clause in each complex sentence. Draw a wavy line under the subordinate clause. Draw a box around the signal word. Add a comma, if necessary, to punctuate the sentence correctly.

Example:

> [If] a small forward bias is applied to the emitter–base junction, the junction barrier will be eliminated.

1. As basic biasing principles state a potential difference across a *PN* junction constitutes a forward bias.
2. When the voltage is connected with negative to the *N*-type material and positive to the *P*-type material, the junction has forward bias.
3. When the polarity is reversed, the junction has reverse bias.
4. A transistor conducts when the emitter–base junction is forward biased and the collector–base junction is reverse biased.
5. Because it has many desirable characteristics the common-emitter configuration is employed more often than any other circuit arrangement.
6. The charges of either *P*- or *N*-type semiconductors can be made to move when an applied voltage produces current.
7. After it is doped, an *N*-type semiconductor has a large supply of free electrons.
8. The reverse current increases with higher temperatures as more minority charges are produced by an increase of thermal energy.
9. Because the reverse current of minority charges is increased, temperature is very important in the operation of *NPN* and *PNP* transistors.
10. Hole current will be produced when voltage is applied across the semiconductor.

Exercise 4 Rewrite the sentences in the exercise above. Reverse the order of the clauses, and punctuate the new sentences correctly.

Example: The junction barrier will be eliminated if a small forward bias is applied to the emitter–base junction.

1. _____

2. _____

3. _____

Complex Sentences

4. _____

5. _____

6. _____

7. _____

8. _____

9. _____

10. _____

Exercise 5 Use the given signal word to combine the two simple sentences into a complex sentence. Add commas if needed. Make sure that the new sentence is smooth and accurate. Write some of the sentences so that the subordinate clause introduces the main clause.

Example: We say that the load is energized.
The current begins flowing through the load. WHEN

When the current begins flowing through the load, we say that the load is energized.

1. Momentary-contact switches are manufactured.
   They always return to the same position after being released.   SO THAT

   _____

2. No electrical connection exists from one terminal of the switch to the other.
   The switch is open.   WHEN

   _____

3. A lamp becomes extinguished.
   The switch is opened.   IF

   _____

4. There are many mechanical designs for switches.
   Different uses require different constructions.   SINCE

   _____

5. A three-way switch is really a single-pole, double-throw switch.
   A three-way switch operates a light fxture from two different locations.   BECAUSE

   _____

6. Cam-operated switches are sometimes called limit switches.
   Cam-operated switches are often used to signal a mechanical motion to stop. BECAUSE

   _____
   _____

7. The toggle switch stays in the position.
   The toggle switch was last placed in a position. THAT

   _____
   _____

8. A pushbutton moves toward the body of the switch.
   The operator presses the button. AFTER

   _____
   _____

9. A built-in spring returns the bar to its normal position.
   The operator releases the button. WHEN

   _____
   _____

10. The finger of a rotary-tap switch makes electrical contact with successive terminals.
    A central shaft rotates the finger. AS

    _____
    _____

THAT, WHICH, AND WHO CLAUSES

Another type of subordinate clause begins with *that, which,* or *who*. These clauses are similar to adjectives because they describe or modify a noun or pronoun in the main clause. In subordinate clauses, these words are both the **signal word** and the **subject** of the subordinate clause.

> Solar Max carried seven scientific instruments *THAT are simultaneously used to monitor a flare.*
>
> These modules, *WHICH are used for such things as power and positioning,* are located in the bottom portion.
>
> The leader of the Astronaut Support Team is Dr. James van Hoften, *WHO is a specialist in shuttle guidance, navigation, and flight control.*

Remember to use WHO and THAT to refer to people and WHICH and THAT to refer to objects. They should be placed directly after the noun or pronoun they modify. Notice in the following two sentences how the meaning changes because of the placement of the dependent clause.

> The team member *that is responsible for safety* is an expert in the shuttle guidance procedure.
>
> The team member is an expert in the shuttle guidance procedure *that is responsible for safety.*
>
> A clause beginning with who, what, or which should be placed as close as possible to the noun to which it refers.

Complex Sentences

Remember, also, that the signal word is similar to a pronoun which refers to the noun in front of it. Even though the signal word is not exactly singular or plural, the *verb* following THAT, WHICH, or WHO must agree with the noun that the clause modifies.

> The team member is an expert in the shuttle guidance *procedure* that *is* responsible for safety.
>
> The team member is an expert in the shuttle guidance *procedures* that *are* responsible for safety.

Exercise 6 Combine the simple sentences into complex sentences using THAT, WHICH, or WHO. Make sure that the new sentence is smooth.

Example: Damage had been done.
Damage caused the satellite to spin unpredictably.
Model: Damage had been done that caused the satellite to spin unpredictably.

1. Nelson returned to the *Challenger*.
   The *Challenger* was 10 minutes away.

2. Four attempts were made to grab the satellite's pin.
   The pin was protruding from the Solar Max.

3. The failed attempts put the satellite into an uncontrolled spin.
   The spin had to be stabilized.

4. The shuttle was carefully positioned by Robert Crippen.
   Robert Crippen was assisted by astronaut Terry Hart in making a successful catch.

5. The satellite's orbit swung around and pointed the panels toward the sun.
   The sun would recharge Solar Max's batteries.

6. The Solar Max carried seven scientific instruments.
   The instruments are all used to monitor a flare.

7. After nine successful months in space, three fuses blew.
   The blown fuses caused the satellite to lose its ability.

8. Drs. van Hoften and Nelson were the space technicians.
   They replaced the defective units on the Solar Max.

9. New satellites will make increasing use of modular construction.
   Modular construction was used on the Solar Max.

   _____

   _____

10. The Solar Max mission was a successful step.
    The step showed that such repair missions are possible and practical.

    _____

    _____

Check your complex sentences with your instructor.

COMPOUND-COMPLEX SENTENCES

Writers vary their sentence structure by using simple, compound, complex, and compound-complex sentences. A **compound-complex sentence** has at least one dependent clause and two or more independent clauses. Notice the combinations and punctuation in the following examples:

> As the frequency is increased, the gain of the amplifier
> *(dependent clause)   (independent clause)*
> is increased, and then it remains relatively constant.
>   *(independent clause)*
>
> It will be found that practical amplifiers, amplifiers
> *(independent clause)   (dependent clause, interrupted)*
> found in actual circuit operation, do not give constant
>   *(dependent clause, con't)*
> gain to all input signals, nor can they withstand all sorts
>   *(independent clause)*
> of different input amplitudes without distortion.

Exercise 7 Combine each group of sentences to make one compound-complex sentence. Be sure to punctuate correctly.

1. a. Consumers keep buying electronic video and audio gadgets.
   b. The number of hand-held remote-control units can create clutter.
   c. Some manufacturers are hoping to profit from a new universal remote controller.

   _____

   _____

2. a. The marketing target is the upscale consumer.
   b. This consumer owns $2000 to $10,000 worth of electronics gear.
   c. The companies believe this consumer will be willing to pay up to $200 for a controller that runs any TV or audio system.

   _____

   _____

3. a. Several vendors have unveiled units.
   b. The units are compatible with different makes of equipment.
   c. Other units are not that sophisticated.
   d. These units are far less expensive—as cheap as $50.

   _____

   _____

4. a. CORE is a unit that can be programmed to run any consumer video or audio electronic system equipped for infrared remote control.
   b. CORE is produced by a company called CL9.
   c. CORE is built around two microprocessors and has only 17 function-control keys.

   _____
   _____

5. a. New units from Zenith and Magnavox will handle only video components.
   b. IR remote is more common with video that with audio equipment.
   c. Others will go after the multibrand video and audio remote-controller market.

   _____
   _____

GRAMMAR UNIT FIVE

SUBJECT/VERB AGREEMENT

Every complete sentence has a subject and a verb. The subjects can be plural or singular, and verbs have tenses. Another consideration in writing standard, complete sentences is subject/verb agreement.

Agreement is a concept similar to that used when matching the right battery to an object. A car battery does not "agree with" a wristwatch. We decide which type of battery we need for a certain object, such as a car or wristwatch, and use only the correct type. In writing, we use the type of verb that agrees with its subject.

Subject/verb agreement is a troublesome problem for some people. We might expect that subjects and verbs follow the same singular and plural rules. They do not—but the rules are easy to follow once they are analyzed.

One comforting fact is that subject/verb agreement is necessary only when writing in the present tense. Also, agreement does not seem to be a problem when the subject is "I" or "you."

This problem centers on the letter S. Read the following sentences.

> The *laser excites* electrons.
> The *lasers excite* electrons.

We usually write the subject first and then make the verb agree with it. For this situation, let's call the plural form of the noun the **S form** (which usually has an S at the end). Verbs also have S forms; however, the rule for when to add an S to a verb is the opposite of the rule for nouns.

Follow these general rules:

──────────── **RULE** ────────────

Add an S to a verb when the subject is singular.
Do not add an S to a verb when the subject is plural.

Or stated another way:

Either the subject or the verb will have an S.

The laser excites electrons.
 (S form)
The lasers excite electrons.
 (S form)

--------------- **RULE** ---------------

In most present-tense sentences, either the subject or the verb will have an "S," but not both.

As we know, some plural nouns do not end in S (men, women, and children), but we consider them as S forms anyway.

The men supervise the test.
 (S form)
The man supervises the test.
 (S form)

Similarly, some nouns always end in S (communications, physics, news, and scissors). In these cases, determine whether you would refer to the noun as *it* or *they*. "Communications" is considered a singular noun ("it" is changing rapidly), and "scissors" is considered a plural noun ("they" are in my desk). Then use the correct verb.

--------------- **WARNING** ---------------

Use the standard spelling rules for adding an S to a verb:
 carry—carries   display—displays

Changing the verb to the past tense eliminates the agreement problem.

The engineer *supervised* the test.
The engineers *supervised* the test.

Changing into the present (-ing) or past (-ed) participle form does not eliminate the problem, since helping verbs also have S forms.

The men WERE supervising the test.
 (S form)
The man WAS supervised during the test.
 (S form)

Notice the S forms of the following verbs.

S Form	Non-S Form
is	are
was	were
has	have
does	do

Exercise 1 Rewrite the following sentences changing the subject from singular to plural, or plural to singular. Then change the verb to agree with the subject. Keep the verbs in the present tense.

Example: Circuits are paths for current.
          (S form)
Changed: A circuit is a path for current.
          (S form)

1. Tiny semiconductor lasers are the heart of grocery store scanners. _____

2. A scientist is working to develop a computer memory that uses a laser.

3. The technique increases computer memory and provides faster access to it.

4. Each card is etched on a thermoplastic square. _____

5. The laser has worked into many facets of society. _____

6. They have transformed the science of eye surgery. _____

7. The laser has been used to keep tunnels true. _____

8. Laser disks store and play back up to 100,000 images. _____

9. An experimental computer memory uses laser disks to store information.

10. A laser processes the images for robots and missile guidance systems.

Special Case 1 Sometimes a sentence begins with an indefinite word such as *there*, *here*, or *where*. In sentences like this, the subject will come after the verb, and the verb must agree with what comes after it.

   Example: There is an illustration.
             There are two illustrations.

_____ **NOTE** _____

The contraction for "there is" is "there's."
"There are" cannot be contracted.

Subject/Verb Agreement

Exercise 2   Using IS or ARE, select the correct form for each sentence.

1. There _____ a similar rule.
2. Where _____ the executives?
3. Here _____ the reason.
4. There _____ three ways of determining power.
5. What _____ the source of power?

Special Case 2   Compound subjects joined by AND are considered plural.

>   The scientist and the researcher agree.

In sentences with subjects joined by OR or NOR, the verb will agree with the closer subject.

>   Neither the scientist nor the *researchers agree.*
>   Neither the researchers nor the *scientist agrees.*

Exercise 3   Underline the correct verb.

1. The laser and the vacuum tube (has/have) been significant developments.
2. Neither prior tests nor this test (is/are) surprising.
3. Either a laser disk or several hard disks (is/are) used in automated teaching systems.
4. Fiber optics and laser communications (has/have) progressed incredibly fast.
5. Neither the sophisticated equipment nor increasing development costs (has/have) slowed down the research.

Special Case 3   When the verb is part of a negative contraction, the verb must still follow the S form rule.

S Form	Non-S Form
(he) doesn't	(they) don't
hasn't	haven't
isn't	aren't

Example:

>   He doesn't like commuting to work.
>   (S form)
>
>   They don't like commuting to work.
>   (S form)

Exercise 4   Underline the correct verb contraction.

1. Lasers still (hasn't/haven't) worked their way into every facet of society.
2. They (doesn't/don't) know of any concept that has had a bigger impact on society.
3. It (doesn't/don't) have many limitations.
4. Many applications (hasn't/haven't) become popular yet.

5. There (isn't/aren't) many technologies developing as quickly as the laser.

Note: There is no CORRECT way of contracting "am not."

SENTENCE COMBINING
This exercise will give you more practice in forming efficient sentences.

Exercise 5 Using the example as a model, combine each group of sentences into a single sentence.

Example: LASER
   a. The laser beam is light.
   b. The light is a special kind.
   c. The light is intense.
   d. It differs from ordinary light.

Combination: The laser beam is a special kind of intense light that differs from ordinary light.

1. a. Ordinary light has many colors.
   b. Its waves move.
   c. The movement is random.
   d. The movement is in many directions.

   Combination: _____

2. a. Laser light is monochromatic.
   b. Its waves move.
   c. The movement is in a single direction.

   Combination: _____

3. a. Photons in ordinary light spread out.
   b. They diffuse energy.
   c. Laser beam photons are concentrated.
   d. They focus their energy.

   Combination: _____

4. a. The beam is generated by a device.
   b. The beam is amplified by a device.
   c. The device is called a laser.
   d. The device emits light waves.
   e. The light waves are spaced.
   f. The light waves are parallel.

   Combination: _____

5. a. The beam has characteristics.
   b. The characteristics are unique.
   c. The characteristics can lead to applications.
   d. The applications can be peaceful.
   e. The applications can be destructive.

   Combination: _____

Subject/Verb Agreement

6. a. A laser can be focused.
   b. It can be transmitted.
   c. It can heat a TV dinner.
   d. The dinner can be a thousand miles away.

   Combination: _____

7. a. But a laser could be focused on a tank.
   b. It could be focused on a ship.
   c. It could be focused on an aircraft.
   d. It could burn a hole.
   e. The burning would be almost instantaneous.
   f. The burning would be through armor.
   g. The armor would be metal.

   Combination: _____

8. a. A laser beam can be used to weld retinas.
   b. The retinas have loosened from the eye.
   c. The welding does not destroy the tissue.

   Combination: _____

9. a. But a laser could be used to destroy.
   b. Human life would be destroyed.
   c. The destruction would be in a flash.

   Combination: _____

10. a. A laser beam can be used to carry signals.
    b. The signals are for radio.
    c. The signals are for television.
    d. The signals are for communication.
    e. The communications are by telephone.

    Combination: _____

11. a. But a laser could also be used from space.
    b. The use would be as a ray.
    c. The ray would cause death.
    d. The death would be to enemies.

    Combination: _____

12. a. Thus laser light is not ordinary.
    b. Its uses are not ordinary.

    Combination: _____

Check your sentences with your instructor.

"Laser" is from *Sentence Combining: A Composing Book,* by William Strong. Copyright © 1973 by Random House, Inc. Reprinted by permission of the publisher.

GRAMMAR UNIT SIX

PREPOSITIONAL PHRASES

Prepositions are words that show relationships between two or more words. They are different from other types of words, and sometimes they are difficult to find.

There are about 30 to 40 common prepositions, some of which are combinations of words. To get an idea of how to identify them, consider the picture of a multimeter and a circuit board.

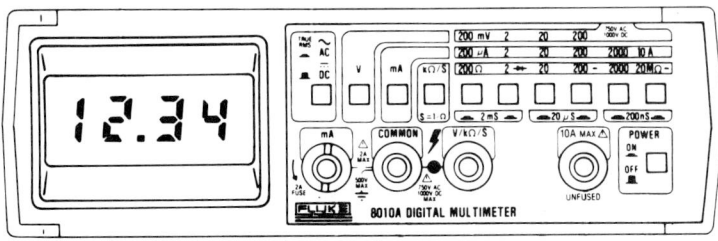

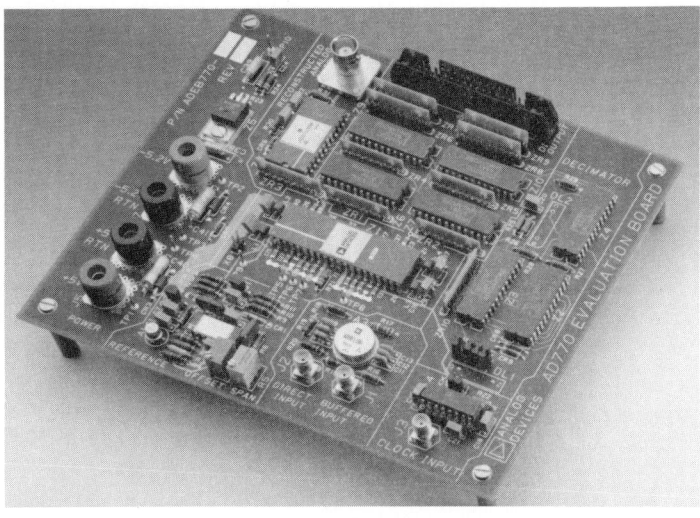

Multimeter control panel courtesy of John Fluke Manufacturing Company. Photo of circuit board courtesy of Analog Devices, Inc.

235

Prepositional Phrases

Think of all the different words you can use to tell where the multimeter is in relation to the circuit board. The words that tell where the multimeter is in relation to the board (on, above, next to) are prepositions.

Common Prepositions

in	into	beyond	in front of
on	onto	among	in back of
at	from	between	in between
to	with	around	beside
by	within	after	next to
of	amid	before	out of
for	through	above	about
		below	

A phrase is more than one word. A **prepositional phrase** begins with a preposition and ends with the next noun (or pronoun) or series of nouns (or pronouns). There could be any number of modifiers inside the phrase.

in the process
for each full-adder

—————— NOTE ——————

The noun at the end of a prepositional phrase is *never* the subject of a sentence.

Exercise 1 Put parentheses around each prepositional phrase. The number at the end of each sentence tells you how many phrases are in that sentence.

1. Noise has recently become a topic of great interest. (1)
2. Legislative bodies have responded to public awareness of the detrimental effects of noise. (3)
3. All communication, with the possible exception of divine revelation, takes place in the presence of noise. (4)
4. For this talk, we will take the usual physics definition of noise. (2)
5. In written communication, noise can occur at each of three elements. (3)
6. The writer can contribute linguistic noise at his semantic encoder. (1)
7. The author must make a selection of data and arguments, but the selection must correspond to the mass of evidence. (3)
8. The commonest source of noise is not the willful distortion of intent, but innocent lapses of attention. (3)
9. Our daily work abounds with good examples of bad habits, all sources of noise. (3)
10. An author who writes with not misspelled words has already dismissed the first grounds of suspicion of the limits of his scholarship. (4)

PREPOSITIONAL PHRASES AND SUBJECT/VERB AGREEMENT

Sometimes a prepositional phrase will occur between the subject and the verb. *The subject will not be inside a prepositional phrase.*

Wrong: The leaders of our *company wants* to expand.
Correct: The *leaders* (of our company) *want* to expand.

Remember to make the verb agree with the real subject, not a noun in a prepositional phrase. Placing parentheses around the prepositional phrase, or drawing a line through it, will help you see the real subject of the sentence.

>   Example: The *values* (of power) *are measured* in watts.
>   The *value* (of power) *is measured* in watts.

Exercise 2 Underline the subject of each sentence once. Choose the correct verb by underlining it twice. Put parentheses around phrases that are between the subject and verb.

1. The rate of energy consumption (depends/depend) on the rate at which power is used.
2. Amounts of power (is/are) predictable for certain uses.
3. The formulas for power (is/are) used for different problems.
4. The surface area of resistors (determines/determine) the amount of power it can handle.
5. The change in the flow of electrons (corresponds/correspond) to a voltage drop.

Exercise 3 Rewrite the sentences in Exercise 2, changing the subject from plural to singular or singular to plural. Then use the correct verb.

1. _____

2. _____

3. _____

4. _____

5. _____

SENTENCE COMBINING

Now practice putting sentences together again, using all the ideas in each group to make one, powerful sentence.

Exercise 4 Using the example below as a model, combine each group of sentences into a single sentence.

TURNING DEFEAT INTO VICTORY

1. a. A friend is a management consultant.
   b. He is successful.
   c. The success is exceptional.
   d. He is my close friend.
   Example: A close friend of mine is an exceptionally successful management consultant.

2. a. You can walk into his office.
   b. You have a feeling.
   c. The feeling is of being "uptown."
   Combination: _____
   _____

Prepositional Phrases

3. a. The furniture is fine.
   b. The furniture tells you something.
   c. The carpeting tells you something.
   d. The people are busy.
   e. The people tell you something.
   f. They tell you his company is prosperous.

   Combination: _____

4. a. A critic would say something.
   b. A critic would call him a man.
   c. The man is a "con."
   d. The critic would be wrong.

   Combination: _____

5. a. The man was not brilliant.
   b. The man was not wealthy.
   c. The man was not lucky.
   d. The man was persistent.
   e. The man never thought he was defeated.

   Combination: _____

6. a. Behind the company is a story.
   b. The company is prosperous.
   c. The company is respected.
   d. The story is of a man.
   e. The man is fighting.
   f. The man is battling his way upward.

   Combination: _____

7. a. In his first six months things happened.
   b. He lost his savings.
   c. The savings were from 10 years.
   d. He lived in his office.
   e. He lacked money to pay rent on an apartment.
   f. He turned down numerous job offers.

   Combination: _____

8. a. He wanted something more.
   b. He wanted to stay with his idea.
   c. He wanted to make it work.
   d. It meant hearing prospects say no.
   e. It happened 100 times as often as they said yes.

   Combination: _____

9. a. It took seven years.
   b. The years were unbelievable.
   c. The years were hard.
   d. During that time, I never heard something.
   e. I never heard my friend complain.

Combination: _____

10. a. He would explain something.
    b. He explained it to me.
    c. He was learning.
    d. The learning was how to sell.
    e. The selling was in a business.
    f. The business was competitive.

Combination: _____

GRAMMAR UNIT SEVEN
PRONOUNS

Pronouns are words that take the place of nouns. Used carefully, pronouns provide a graceful substitution for repeated nouns. Even though the dictionary contains thousands of nouns, we have only a small number of pronouns, and each pronoun has a specific function in a sentence. There are four main groups of pronouns: subject, object, possessive, and reflexive.

SUBJECT PRONOUNS

Subject pronouns are used to replace the subject of a sentence. The subject pronoun must agree with the "voice" and "number" of the subject. The subject pronouns are listed below:

	Singular	Plural
First person	I	we
Second person	you	you
Third person	he, she it, one	they
Question form	who	who

Exercise 1 Underline the subject/s in each sentence. Write the pronoun that would replace the subject.

1. _____ Science covers the broad field of human knowledge.
2. _____ Scientists discover and test facts and principles by using the scientific method.
3. _____ Nikola Tesla pioneered the development of radio and high-temperature electricity.
4. _____ Electronics is a branch of physical science.
5. _____ Which people are on the research team?

HAZARD

Pronouns should not be used as the subject in the first sentence of a paragraph. A subject pronoun should only be used after the noun has been introduced in each paragraph.

OBJECT PRONOUNS

Another group of pronouns is used to replace nouns that are objects, such as those in prepositional phrases or after verbs.

	Singular	Plural
First person	(to) me	(to) us
Second person	you	you
Third person	him, her it, one	them
Question form	whom	whom

Exercise 2   Replace the italicized object nouns with a pronoun.

Example: (US)   Mr. Reeder gave the broken component to *our team*.

1. _____ First we tried to isolate the cause of the problem, and then we tried to fix *the problem*.
2. _____ We were not sure to *which people* we could turn for advice.
3. _____ After locating the special tools, we used *the tools* to find the faulty unit.
4. _____ The technical manuals that we borrowed from *Mr. Reeder* contained schematic diagrams.
5. _____ The results were sent back to the *development personnel* for further evaluation.

COMPOUND PRONOUNS

If more than one pronoun is used as a subject or object, be sure that each pronoun is the correct form for the situation. When deciding which form to use, try each pronoun individually. Always refer to yourself last.

   Example: He and I arrived at the same conclusion.
            HE is the subject form—HE arrived.
            I is the subject form—I arrived.
            The I pronoun comes last in the combination.

   Example: The prize money was shared by them and us.
            THEM is the object form—by THEM.
            US is the object form—by US.
            The US pronoun comes last in the combination.

If a pronoun is used with nouns, be careful to use the correct form of the pronoun for the situation.

   Give the results to Chris and me.
   Chris and I will evaluate the data.

Exercise 3 Circle the correct pronoun in each sentence.

1. Tim, Angela, and (I/me) are lab partners.
2. Tim didn't do as much work as Angela and (I/me) did.
3. We showed our data to Bill, Nick, and (he/him).
4. Our instructor gave a warning to (he/him) and (we/us).
5. He said that Angela and (I/me) could no longer share results with (they/them) and Tim.

POSSESSIVE PRONOUNS

This group of pronouns shows ownership. The choice of **possessive pronouns** varies depending on whether it is followed by a noun or whether the possessive pronoun is used by itself.

Example: That is *my* hypothesis.
The hypothesis is *mine*.

	Singular	Plural
First person	my, mine	our, ours
Second person	your, yours	your, yours
Third person	his her, hers its, one's	their, theirs
Question form	whose	whose

Exercise 4 Circle the correct possessive pronoun in each sentence.

1. (You/your) experience in troubleshooting is limited, but (their/theirs) goes back many years.
2. Each scientific field regulates (it/its) own professional standards.
3. The choice is (your/yours).
4. I had to make up (my/mine) own mind.
5. The team leader claimed that the responsibility was (her/hers).

——————— HAZARD ———————

Do not put apostrophes in possessive pronouns except in the impersonal pronoun, *one*.

Example: A career choice is one's own decision.

In other cases, adding an apostrophe changes the function of the word.

It's stands for *it is*.
You're stands for *you are*.
Who's stands for *who is*.

REFLEXIVE PRONOUNS

Reflexive pronouns refer back to someone or something. They end in *self* or *selves*. They are used sometimes to add emphasis to the person or people already named in the sentence.

Example: Check the results yourself.
(Referring to the understood "you")

	Singular	Plural
First person	myself	ourselves
Second person	yourself	yourselves
Third person	himself	themselves
	herself	
	itself	
	oneself	

Do not use reflexive pronouns if the noun has not been used in the same sentence.

Wrong: The results were checked by myself.
Problem—"myself" has no person to refer to in the sentence.
Correct: The results were checked by *me. (using object pronoun)*
Correct: *I* checked the results. *(using the subject pronoun)*

──────────── **HAZARD** ────────────

There are no such words as *hisself, ourself, theirself, themself, selfs,* or *theirselves.* Be careful to spell pronouns correctly.

Exercise 5 Fill in the correct reflexive pronoun.

1. He cheated _____ out of a valuable learning experience by not completing the experiment.
2. You must discipline _____ to be systematic and logical.
3. It is one's responsibility to teach _____.
4. The determined technicians trained _____ to use the six-step troubleshooting method.
5. We could not bring _____ to admit that most of our tests were unnecessary.

Special Case 1: Nouns Ending in "S." Some nouns always end with a final "S." Some of them are considered singular and are replaced by a singular pronoun. Others are considered plural. Use your ear to decide which pronoun is right.

Singular Nouns (pronoun: IT)	Plural Nouns (pronoun: THEY)
communications	eyeglasses
fiber optics	pants
physics	scissors
electronics	pliers

Special Case 2: Indefinite Pronouns. Indefinite pronouns are always considered singular. Some indefinite pronouns (everybody, everything) may actually be referring to several people or things, but each person or thing is being referred to individually. The following list shows several of these singular pronouns.

244   *Pronouns*

anyone	anything	anybody
everyone	everything	everybody
no one	nothing	nobody
someone	something	somebody

Indefinite pronouns as subjects of sentences require S-form verbs (in the present tense) and S-form helping verbs.

Example: Everyone is invited to attend. (every *one* person)
No one has completed the assignment. (not *one* person)
Everything is finished. (every *one* thing)

Exercise 6   Circle the correct verb in each sentence.

1. Everyone (has/have) finished the experiment.
2. My pliers (was/were) left on the lab bench.
3. Someone (was/were) going to ask the instructor.
4. Until now, mathematics (has/have) been my easiest subject.
5. Everything about this experiment (was/were) confusing.

Exercise 7   Write a complete sentence that uses the following pronouns correctly.

1. *their* _____
2. *mine* _____
3. *themselves* _____
4. *he* and *his boss* _____
5. *who* _____
6. *hers* _____
7. *one's* _____
8. *itself* _____
9. *you* and *her* _____
10. *everybody* _____

GRAMMAR UNIT EIGHT

PRONOUN REFERENCE

Pronouns are singular or plural, and they have subject, object, possessive, and reflexive forms.

Exercise 1 As a review, fill in the table of pronouns.

	Singular				*Plural*			
	Subject	Object	Poss.	Reflex.	Subject	Object	Poss.	Reflex.
1st	I	_____	my/mine	myself	_____	us	our/s	_____
2nd	_____	you	_____	yourself	you	_____	your/s	_____
3rd	he	_____	his	himself	they	_____	_____	themselves
	she	her	_____	_____				
	it	it	_____	_____				
	one	one	_____	_____				

Every noun can be referred to by a pronoun.

>   Dr. Thompson = *she/he*
>   gentleman   = *he*
>   dog   = *it* (or *he, she*)
>   players   = *they*
>   resistor   = *it*
>   resistors   = *they*

Each pronoun must refer to its noun in the same "number" (singular or plural) as the noun is last written. If the noun is plural, the pronoun referring to it must be in the plural form, also.

The *lady* asked where *she* could find the library.
Tom and Larry found *they* needed more information.

Exercise 2 Circle the correct pronoun. Draw an arrow to the noun that the pronoun refers to. Circle the noun, also. The first one is done for you.

1. Put the (tools) back in (its/(their)) compartments.
2. The research department met all (its/their) deadlines for the month.
3. The scientist applied for (his/their) first patent.
4. Ted asked the guests to make (theirselves/themselves) at home.
5. We tried to finish the experiment (ourself/ourselves).
6. The student grabbed (his/one's) notebooks as he ran out the door and put (it/them) in (his/their) briefcase.
7. The students complained that Dr. Johnson spoke too quickly to (him/them).
8. Ohm's law is useful because (it/they) shows the relationship between current, voltage, and resistance.
9. Resistance, current, and voltage must be measured in (its/their) individual types of units.
10. Voltages can be added as long as (it/they) are connected with series-aiding polarities.

Collective nouns are words that have a singular form but may be singular or plural in meaning. Some examples of collective nouns are *team, group, class, audience,* and *series*. When we use a collective noun, we are referring to the "collection" as one unit, and we refer to it with a singular pronoun.

The *band* played *its* theme song.
The *research team* completed *its* final report.

However, if we are referring to individual members of the unit, we use a plural pronoun.

The *members* of the band played *their* theme song.
Most of the team turned in *their* results.

Exercise 3 Circle the correct pronoun. Draw an arrow to the noun to which it refers.

1. A team was selected in January, but (it/they) did not meet until the following June.
2. The technicians on the team received (its/their) instructions from the team leader.
3. The team presented (its/their) report at the end of the year.
4. This group of resistors is known for (its/their) durability.
5. A visiting group of scientists from Japan finished (its/their) tour of the factory before meeting with the president.
6. Two people in the group discussed (his/their) observations with the chief engineer.

7. The class of 1980 held (its/their) reunion at the college.
8. Four people from the class sent (his/their) regrets.
9. One class of technician trainees became well known for (its/their) final projects.
10. The first series of tests was completed before (its/their) deadline.

GRAMMAR UNIT NINE

MODIFIERS

Modifiers are words that limit and describe other words. We use modifiers in writing to add specific details to objects, people, feelings, and actions. Compare the following descriptions:

> The man wearing jeans, a shirt, and shoes walked into the lab.
> The bearded young man wearing faded Levis, a red Nike T-shirt, and high-top sneakers confidently walked into the fast-paced development lab.

The two sentences have the same basic idea. But the first sentence sounds flat and boring when compared to the second sentence, which is full of modifiers. Used wisely, modifiers add power and life to writing. The most common modifiers are called *adjectives* and *adverbs*.

ADJECTIVES

Adjectives are modifiers that describe nouns or pronouns. Some sentences have no adjectives, whereas other sentences have many.

> The salesperson wore a suit.

This simple sentence has a subject and verb; however, the picture communicated by the writer is brief and incomplete. Two people reading this sentence may have completely different "pictures" of this salesperson. Neither the person nor the suit is described. If description makes a difference, such as in describing a fashionable suit or a certain salesperson, adjectives are needed.

Modifiers 249

The *young* salesperson. . . .
The *tailored* suit. . . .

Although adjectives are usually in front of nouns, another way of writing these same ideas uses linking verbs.

The salesperson is *young*.
The suit is *tailored*.

Supply adjectives of your own:

The _____ salesperson. . . .
The _____ suit. . . .
The new salesperson wore a (an) _____ suit.
The _____ salesperson wore a three-piece suit.

_____ **NOTE** _____

The articles A, AN, and THE are also adjectives since they are written in front of nouns and pronouns.

Exercise 1 Underline the adjectives in the following sentences. The number of adjectives in each sentence is indicated in the parentheses. Draw an arrow to the word each adjective describes.

1. The consultant advised men to wear a blue or gray suit to the first interview. (6)
2. A black suit is overpowering. (3)
3. Black accessories often complement a business suit. (3)
4. A woman has several more color choices for a business suit. (6)
5. Blue is the favorite color of most men and women. (3)
6. Brown is a good, basic color for women, but should only be used as a background or accessory color by men. (6)
7. Burgundy is one of the most flattering, authoritative colors. (4)
8. Women can wear burgundy suits or accessories. (1)
9. Men can wear burgundy ties, either as a background color or a print, but they should not wear burgundy suits or shoes. (5)
10. Gray and navy business suits are the most authoritative colors for men or women. (6)

HYPHENATED ADJECTIVES

Sometimes two words are joined by a hyphen to become one adjective, as in the following examples.

middle-class neighborhood   navy-blue three-piece suit
top-level management   well-polished image

It would sound ridiculous to talk about a "middle neighborhood" or a "class neighborhood." Instead, the two words have been joined with a hyphen to become the single adjective "middle-class."

RULE

If two words require each other to describe the noun, use a hyphen between them.

Exercise 2 Add a hyphen between the adjectives that require each other to describe the noun.

Example: Graduation is a deicsion - making time.

1. He has an easy going attitude about interviews.
2. A lemon colored suit is not appropriate for an interview.
3. Some businesses allow an open collared, button down shirt.
4. The interviewer was impressed by her well polished image.
5. He bought a long sleeved, white shirt.

ADJECTIVES FOR COMPARISON

We often indicate comparisons with adjectives by using MORE or MOST.

> A camel suit is a *popular* choice.
> A blue suit is a *more popular* choice.
>   (comparing blue to camel)
> Gray is the *most popular* choice of all.
>   (comparing gray to all others)

Sometimes we add the suffix -ER or -EST to the end of the adjective.

> The blue tie is bright.
> The blue striped tie is brighter.
>   (comparing blue to striped blue)
> The red tie is the brightest.
>   (comparing red to all others)

RULE

Use more *or* -ER to compare one thing to one other thing.
Use most *or* -EST to compare one thing to all others.

RULES OF THUMB

If the adjective has one syllable, add ER (a cleaner suit).
If the adjective has three or more syllables, use MORE (a more expensive suit).
If the adjective has two syllables, use either ER or MORE (a more useful suit, but an easier choice).

Note It is redundant (and incorrect) to use both *more/-er* or *most/-est*.

> *Wrong:*   This suit is *more bluer* than that suit.
> *Correct:* This suit is *more* blue than that suit.
>
> *Wrong:*   The navy-blue suit is a *more usefuler* choice than black.
> *Correct:* The navy-blue suit is a *more useful* choice than black.

Exercise 3 Compare the following adjectives by choosing the correct comparative form.

Example:
harmful	more harmful	most harmful
	MORE/-ER	MOST/-EST
1. hard		
2. serious		
3. easy		
4. slow		
5. fine		
6. accurate		
7. precise		
8. sharp		
9. correct		
10. complete		

The three common exceptions are *good, less,* and *bad.* The comparative forms of these words use neither -er/more nor -est/most. They use different words. See if you can figure them out.

	ER-form	EST-form
good	better	_____
bad	_____	worst
less	lesser	_____

Check with your instructor or the dictionary if you're not sure.

──────────── **NOTE** ────────────

Some words cannot be used to compare, such as *dead, unique,* and *only.* Each of these simply states a condition.

Wrong: This color looks deader than that color.
Correct: This color looks dead compared to that color.

Wrong: This tie is more unique.
Correct: This tie is unique.

Also, technical writers must be careful about overusing comparative words such as *real, really, nice, pretty good, pretty well,* and *very.* These words do not have a clear meaning and should be replaced by more specific terms.

Wrong: That tie looks really nice.
Correct: That tie looks professional.

Exercise 4 Edit the adjectives in the following sentences to make them correct and clear.

1. Blue looks more better on you than gray.
2. The reception was a really nice party.

3. My manager is a really nice person.
4. That is the baddest-looking outfit I have ever seen.
5. Meeting the president was a most unique experience for me.
6. We had the seriousest conversation of the evening.
7. I think she was more sharp than anyone else at the party.
8. I gave him my most firmest handshake.
9. We were all using our correctest manners.
10. This clock is much more slower than my watch.

ADVERBS

An **adverb** is a word that describes a verb, an adjective, or another adverb. It is used to answer certain questions in a sentence:

How?   When?   Where?   To what degree?

Henry never wears a blazer to an important meeting.
(When? Never!)

He did not fully accept the concept of a "power suit."
(To what degree? Fully!)

The college graduate dressed professionally.
(How? Professionally!)

——————— **NOTE** ———————

Many adverbs end in "LY." We add "ly" to most adjectives to make them function as adverbs.

He wore attractive clothes.
(*adjective* describing the clothes)

He dressed *attractively*.
(*adverb* describing how he dressed)

Exercise 5   Underline the adverb that answers the question after each sentence.

1. He decided to wear the suit again. *(When?)*
2. She relied heavily on her navy-blue suit. *(To what degree?)*
3. When he examined his wardrobe, he realized that he had almost no blue. *(To what degree?)*
4. The manager told him not to wear a three-piece suit here. *(Where?)*
5. Some lab technicians always wear a white jacket. *(When?)*
6. Many professionals pride themselves on dressing well. *(How?)*
7. There is currently an emphasis on certain conservative styles. *(When?)*
8. The expense of a suit means students will shop for clothes carefully. *(How?)*
9. Green hasn't ever been a popular choice for men's suits. *(When?)*
10. Dressing professionally will increase your chances of being successful. *(How?)*

DOUBLE NEGATIVES

In mathematics, two negatives make a positive. In language, two negatives also make a positive. Used carefully, a **double negative** can be used to add emphasis to a point.

"It is not unlikely" means "it is likely."

Used carefully, a double negative can be used to add emphasis to a point. However, some writers use a double negative without intending the meaning to be positive. Most people are aware of the rule against using two negative words for the same idea, but some get careless or lazy about making negative statements correctly.

CAUTION

People who intend to project an educated image will never use "ain't." If you use "ain't," break the habit quickly and substitute *isn't, aren't, haven't,* or *am not.*
Wrong: He ain't dressed like a professional.
Correct: He isn't dressed like a professional.

Some words have a negative meaning, and should not be used with another negative word:

| hardly | no one | rarely |
| never | none | scarcely |

Wrong: *No one scarcely* attends the early labs.
He *never* has *none* of his work finished on time.

Some words have a positive meaning and are often used correctly with negative words:

any   anyone   ever

Correct: *No one ever* attends the early labs.
Hardly anyone attends the early labs.
He *never* has *any* of his work finished on time.

The following examples correct a double negative, each adding a slightly different emphasis.

Wrong: I haven't hardly started thinking about clothes.
Correct: I have hardly started thinking about clothes.
I haven't started thinking about clothes.
I hardly think about clothes.

Wrong: We don't never shop there.
Correct: We never shop there.
We don't ever shop there.
We don't shop there.

Exercise 6 Rewrite the following sentences to correct the double negatives.

1. I haven't never seen a man wear a yellow suit. _____

2. It ain't no proper color for a professional. _____

3. I can't never find the time to shop. _____

4. They don't hardly know where to begin. _____

5. Don't let no one talk you into buying a suit you don't like. _____

Exercise 7 Proofread the following paragraph for modifier and double-negative errors. You should find six errors. Write the correct forms above them.

Our response to color is as much emotional as physical. Certain colors deliver messages that haven't nothing to do with the individual wearing them. More darker colors generally convey authority. Medium-range colors like blue or tan make us look more friendlier and approachablier. Large expanses of pastels make us look less seriouser, sometimes unprofessional, and are more suited for off-the-job looks. Our response to colors may not be hardly deliberate, but colors contribute to our general impression of the person wearing them.

MISPLACED MODIFIERS

A modifer can be a phrase that describes a word in the sentence. Be careful to place the modifying phrase close to the word it describes. **Misplaced modifiers** can be silly, as in the next sentence:

Misplaced: I saw a quarter walking down the street.
(Was the quarter walking?)
Correct: Walking down the street, I saw a quarter.

Some misplaced modifiers, such as ONLY, JUST, and ALMOST, can make the meaning unclear, as in the following sentence:

He just planned the circuit.

This sentence could be interpreted two ways—he planned the circuit but left the construction for someone else, or he planned the circuit recently. Be careful to add enough information for the reader.

Correct: He planned the circuit just now.
Or: His part of the project was planning the circuit.

Exercise 8 Rewrite the following sentences to eliminate misplaced modifiers.

Example: Despite their intensive training, robot failures are sometimes difficult for maintenance technicians. (The robots didn't have the training.)

Revised: Robot failures are sometimes difficult for maintenance technicians, despite their intensive training.

1. A major appliance manufacturer only purchased an industrial robot. _____

2. After it was installed, the company had a series of problems with the robot.

3. The robot was just designed to load and unload a stamping press. _____

4. After breaking down, the maintenance crews pulled the robot off the line and replaced it with a human operator. _____

5. Finally, the robot was pulled away by the supervisor from the stamping press permanently. _____

6. When not properly designed and installed, the factory is a rough environment for a robot. _____

7. Unlike other factory machinery, human beings can quickly replace problematic robots. _____

8. Early failures tended to discourage further investments in robotics technology of time and money. _____

9. Today, experienced consultants can be hired to ease the transitions that are experts at building robotics systems. _____

10. Robotics systems operate far more successfully that are built by systems manufacturers than those built and installed in-house. _____

GRAMMAR UNIT TEN

PARALLELISM

When you first learned about electronic circuitry, you learned the difference between series and parallel circuits. Parallel circuits, you learned, are constructed so that an equal voltage is applied to all components. In geometry, parallel lines never intersect because they are always an equal distance apart. We use a similar concept of parallelism in writing.

Parallelism in writing is a method of using similar words or sentence structures to identify equal ideas. A series of nouns, verbs, or phrases must be expressed in a parallel form. To keep ideas parallel, we state them in similar forms and word orders. Sentences that are not in parallel form sound awkward and confusing.

The problem in the following awkward example lies in the verb tense and word order.

Awkward: Management communication can promote effectiveness or be destroyed.

The following corrected sentence puts the verbs in a parallel form (present tense) and reorders the words to make the idea logical and clear.

Parallel: Management communication can promote or destroy effectiveness.

Putting a sentence into a clear, parallel form often gets rid of clumsy wording and "noise."

Awkward: Even the best communicators learn as much as possible about their listeners and tailor their remarks to their interests, attitudes and what their values are.

Parallel: Even the best communicators learn as much as possible about their listeners and tailor remarks to the listeners' interests, attitudes, and values.

Occasionally, repeating a similar form can add emphasis to the equality of the ideas. The following examples show different types of parallelism.

Parallel Words and Phrases

While all we want is to *sound good*
 look good or
 appear to be intelligent,
some people get carried away.
(all present-tense verbs)

Failing to close the feedback loop can take several forms:
 not listening to our own messages, and
 not listening to our receiver's feedback.
(repeated NOT and -ING forms)

Parallel Clauses

When an unhappy employee bellows, "I'm not mad!" it's pretty obvious that *what he or she is saying and*
 what he or she is communicating
are two different things.
 (repeated WHAT, IS, and -ING verb forms)

Parallel Sentences

Tailor your language to your audience.
Or, when in doubt, *use simpler language.*
 (repeated understood YOU and present-tense verbs)

Exercise 1 Underline the parallel ideas in the following sentences.

1. Remember that what you do speaks louder than what you say.
2. You should tell the reader what is coming up and make it easy for that person.
3. Much of the meaning we convey to other people, we convey through our tone of voice, appearance, timing, and many other nonverbal factors.
4. People will not communicate effectively if they fail to adjust language to their audience or fail to word ideas in an understandable way.
5. Your presentations will be more successful if you state your purpose, define unfamiliar terms, present information logically, and restate important ideas in a conclusion.
6. Transitions in presentations are more easily identified if you state your objectives clearly at the beginning and if you use sequence words, such as "first" and "second," during the talk.
7. Messages that use simple language will be better understood by your co-workers, by your managers, and by your friends and family.

8. There are kernels of information that are important to understand and necessary to convey.
9. In written communication, "access" means using enumeration, white space, short paragraphs, highlighting, and bullets to identify bits of information.
10. No one wants to wade through a confused piece of writing, and no one wants to find a surprise twist at the end.

SUGGESTIONS FOR WRITING PARALLEL IDEAS
The following suggestions will make parallel ideas more clear.

Hint 1 Repeat a preposition, an article, or the introductory word of a phrase or clause.

> *Examples:*
> People ensure effective communication
> > by adjusting language to their audience or
> > by wording ideas in simple ways
>
> They fail to adjust language to their audience, or
> > to word ideas in ways that others can understand.
>
> Two factors lie at the root of this sin:
> > a failure to recognize differing purposes and
> > a desire to exert power.
>
> A message is effective when it
> > is reached by its intended audience,
> > is understood by the receiver,
> > is remembered for a reasonable time, and
> > is used when appropriate.

Hint 2 Use correlatives such as the following:

> both . . . and   not only . . . but also
> either . . . or   whether . . . or
> neither . . . nor

> *Examples:*
> Communication includes both the delivery of a message
> > and the understanding of a message.
>
> Actions not only speak louder than words,
> > but also can confuse or contradict words.

Hint 3 Be sure that items in a series are in a similar form.

> *Examples:*
> The confused listener was filled with questions,
> > frustration, and
> > rage.
> > > (series of nouns)
>
> An effective communicator can present a message that is clear,
> > complete, and
> > concise.
> > > (series of adjectives)

Exercise 2 Rewrite these awkward sentences in a parallel form.

> *Example:* Computer systems called "computer-aided design" (CAD) and "manufacturing" (CAM) are useful for engineers.
>
> *Model:* Computer systems called "computer-aided design" (CAD) and "computer-aided manufacturing" (CAM) are useful for engineers.

1. CAD/CAM systems enable engineers to rapidly design mechanical parts and analyzing them to close tolerances. _____

2. By avoiding work on the shop floor that is time consuming and expensive, companies can reduce the time of developing new products and cost-wise. _____

3. The time between the conceiving and the finished product affects the development cost. _____

4. In a CAD system, computers make calculations for both optimum shapes and to the size for a variety of applications. _____

5. The advantages of a CAD system are its speed and are more accurate in job performance. _____

6. Designers are not only free from the repetitive task of drawing lines but also calculating workpiece sizes is difficult. _____

7. Once developed, CAD drawings can be stored in computer memory either as dynamic, three-dimensional forms, or if they aren't, then the multiview projections are more conventional. _____

8. A CAM system communicates the work instructions for automatic machinery, such as robots, to handle, process, and by producing a product. _____

9. Until recently, the CAD/CAM industry predominantly marketed stand-alone machines, but now four or five workstations operate under the direction of a central computer are the systems that vendors market. _____

10. Computer programs, called software, are the principal form of communication between the programmer and when the computer is used. _____

Exercise 3 The following paragraphs need revision. They have errors due to awkwardness, fragments, and redundancy. Rewrite the paragraphs to make them clear and effective. Revisions will vary.

1. The personal computer is the consummate tool. It can work for you and linked to other computers. When linked to other computers, it has access to a tremendous range of information services to retrieve, store, reorganization, and shipment of data.

260 *Parallelism*

2. People from many professions and interests use computers every day. These people report making more money, they had more fun, or being more powerful as a result of using personal computers. Their success stories started by calling up information utilities that maintain data banks, or they called up other users to ask questions and have gotten immediate answers.

3. Computer communication is a complex topic. There are several things you need to know, or someone else who knows them. If you know someone who already understands computer communication. Just sit back and listen. But if you are on your own. You must learn something about computers on your own.

4. Computers store information in their internal memory or floppy or hard disk, but no one can own all the information he could need. Now databanks can collect, maintain, will purge dated information, and can update all files.

5. Databanks are either narrow, well-defined systems that serve a specific target group. Or more people use multipurpose systems that provide broad types of information. Multipurpose utilities usually perform at least three major functions: communications, allowing transacting, and information access.

GRAMMAR UNIT ELEVEN

AVOIDING SHIFTS

Keeping a consistent tone and voice is important in effective writing. Abrupt changes in tone or, even worse, illogical shifts in voice or number make readers uneasy and confused. The four most common shift errors in writing occur in verbs, pronouns, voice and tone.

VERB AND PRONOUN SHIFTS

Verb shifts can occur whenever two or more verbs are used in a sentence. When the verbs are used to express a single time frame, they must remain in the same tense.

> *Wrong:* The dielectric *does* not pass measurable current unless the field strength *was* strong enough to cause breakdown. *(present to past tense shift)*
>
> *Correct:* The dielectric *does* not pass measurable current unless the field strength *is* strong enough to cause breakdown. *(consistent present tense)*

There are some occasions in which different verb tenses are logical and necessary.

> The dielectric strength *is* a measure of the maximum voltage, *measured* in volts per mil, that *can be applied* without breakdown of the insulator.

Do not use consistent tenses when clarity and logic demand other forms.

Pronoun agreement means using pronouns that agree in number (either singular or plural) with the nouns to which they refer. Remember that pronouns have singular and plural forms whether they are in the subject form (I, we), object form

261

(me, us), possessive form (mine, ours), or reflexive (myself, ourselves). For a review, refer to Grammar Unit Seven.

> Wrong: *Molecules* are said to be polarized, and *it* orients *itself* in accordance with the field.
> Correct: *Molecules* are said to be polarized, and *they* orient *themselves* in accordance with the field.

Exercise 1 Correct the unnecessary verb and pronoun shifts in the following sentences. Cross out any incorrect form and write the correct form above it.

1. Capacitors are manufactured with different capacitance values depending on its dielectric material.
2. Mica capacitors are made of sets of parallel plates that is insulated by thin sheets of mica.
3. These are available in units ranging from a few picofarads to a few nanofarads, making it useful at high radio frequencies.
4. Paper capacitors are made using strips of aluminum foil with treated paper as the dielectric, and the foil and dielectric were rolled into a cylindrical form.
5. This capacitor may be sealed in wax paper or plastic, depending on their applications.
6. Ceramic capacitors use various types of ceramic dielectrics, and it tends to have values of capacitance with large values of tolerance.
7. Electrolytic capacitors have a positive terminal of aluminum foil and another plate is made of aluminum hydroxide, and it is packaged in an aluminum cylinder.
8. This capacitor can be used at dangerously high currents, but if operated with reverse polarity, they overheated and exploded.
9. Variable capacitors are constructed so that its value of capacitance can be varied.
10. These capacitors are used as tuning capacitors in variable-frequency circuits, where its capacitance was adjusted to specific amounts.

VOICE SHIFTS

Voice shifts refer to both the person (first, second, or third) and to active and passive voice. Remember that the active voice shows a subject performing the action of the verb.

The technician calculated the capacitance.

The passive voice shows an inactive subject and the verb includes a helping verb.

The capacitance was calculated by the technician.

Determine your voice before beginning to write. Choose the appropriate voice for your audience and purpose. If your readers need directions or instructions, use the second person, active voice. If your readers need information, use the first or third person (first for informal communications, third for more formal communications), active voice. Then proofread for consistency after the rough draft is finished.

The easiest way to describe "person" is to look at the types of pronouns used in each style. The *first person* uses "I" and "me" because the writer is describing personal experiences.

> I placed the clips on the coil.

The *second person* uses "you" or the "inferred you" because the writer is directing the reader.

> Place the clips on the coil.

The *third person* uses "he," "she," "it," or "they" because the writer is informing the reader of someone or something else.

> They placed the clips on the coil.

The most common voice shift error is using YOU after starting with another pronoun.

> *Wrong:* I found that superposition works well when you need to simplify a circuit.
> *Correct:* I found that superposition works well when I need to simplify a circuit.
> *Correct:* I found that superposition works well for simplifying circuits.

Hint 1 Use the first person (I, me), active voice for memos, letters, and lab reports.

> *Wrong:* I found that the total capacitance was less than was found in the smallest capacitor.
> *Correct:* I found that the total capacitance was less than I found in the smallest capacitor.

Hint 2 Use the second person (you or inferred you), active voice for directions.

> *Wrong:* Be careful when connecting capacitors in series that we have not exceeded the voltage rating of the capacitor.
> *Correct:* Be careful when connecting capacitors in series that you do not exceed the voltage rating of the capacitor.

Hint 3 Use the third person (it, they), passive voice for formal types of writing, such as research papers or formal proposals.

> *Wrong:* The total capacitance was determined to be less than I found in the smallest capacitor.
> *Correct:* The total capacitance was determined to be less than that of the smallest capacitor.

Exercise 2 Edit the following paragraphs to eliminate voice shifts. Circle the shifts and add the correct forms.

1. When you begin your job search, it is important for one to avoid making several common mistakes. Any one of these mistakes can ruin his chance of getting the job you want.

2. The first error is not knowing what you want to do. Employers don't want to hear, "What's available? You'll do anything!" A qualified and motivated person is one whom the employer wants.

3. Job candidates must take the initiative to obtain your ideal job. They can start in small ways, such as making lists of steps or discussing job-hunting strategies with a professional. Consider the job hunt your immediate job.

4. Another mistake is going to too few prospects. Some new job hunters approach only a few leads at a time. I was discouraged when those leads didn't work out, and felt like I had to start the search over. Job hunters who only contact a few leads seem to take rejections more personally.

5. You must be able to view employment from the employer's perspective. I have reasons for a job, but they are of no concern to the potential employer. One must be able to focus on the needs, objectives, and problems of the employer.

Revisions may vary.

TONE SHIFTS

Tone can most easily be described as the attitude of the writer. The tone can be formal or casual, commanding or friendly, funny or angry. Shifting from one tone to another is confusing to readers. They may feel manipulated or misled. This problem in writing is the hardest to identify and correct because it does not consist of grammatical errors. It consists of word choice and phrasing.

Words have a context, or "tone" about them. For example, "trim" and "skinny" have the same general meaning, but if you were describing a woman's figure, "trim" would be taken as a compliment, while "skinny" would more likely be taken as an insult. Read the following pairs of synonyms, and determine which column has a more positive tone, and which a more negative tone.

Synonyms with different tones:

suggest	insinuate
rumor	gossip
assertive	aggressive
pause	hesitate
eager	brash
enthusiastic	fanatical
knowledgeable	know-it-all

The words in the first column are considered complimentary. The words in the second column, although they have the same general meaning, are considered more negative in tone.

Some words or phrases are considered "loaded," which means that they have strong emotional meanings. Examples of loaded words are listed below.

Loaded words:

red-neck	hippie	yuppie
hot-head	sexy	leading edge
kook	excellent	state-of-the-art
lean and mean	macho	feminist

Other words with loaded meanings include nicknames for ethnic, religious, or political groups; political sayings; and sexual terms. If a loaded word or phrase is insulting, it is not appropriate in any type of professional writing. Some loaded words are harmless or possibly useful. For example, a "lean and mean" department currently implies a small, highly trained staff capable of working intensively on projects. However, some expressions have become overused and have lost a clear meaning, such as "a nominal fee." Overused or trite expressions should be avoided.

Some words and expressions are considered formal, neutral, or slang. In the following list, compare the synonyms that set a formal, neutral, or slang tone.

Formal Words	Neutral	Slang
fortuitous	fortunate	lucky
contemplate	consider	chew on
copious	many	gobs
reiterate	repeat	ditto
elucidate	explain	draw you a picture
dialogue	conversation	rap
recalcitrant	stubborn	muleheaded
disconcerting	upsetting	a downer

Try to keep a neutral tone in most business writing. Avoid abrupt shifts in tone, as in the example below.

> I am interested in applying to your company for the position of lab technician. I have this brainy friend who works for you and he thinks your company is proliferating and in tip-top shape.

The revised paragraph eliminates the loaded word (*brainy*), slang (*tip-top*), and overly formal word (*proliferating*).

> I am interested in applying to your company for the position of lab technician. A friend who works for you has described your company as productive and well managed.

Exercise 3 Decode the following common expressions which have been garbled with overly formal words. Write the common expression.

Example: Dispatch remuneration expeditiously.
Decoded: Send money soon.

1. Nothing flourishes like prosperity. _____

2. The fourth dimension is a medium of exchange. _____

3. Bestir yourself and detect the redolence of java. _____

4. Rapidity of motion makes refuse and debris. _____

5. Dual craniums are superior to sole examples. _____

6. Operational deeds converse with more intensity than units of a declaration. _____

7. If you can't subjugate them, enroll. _____

Avoiding Shifts

8. Allow lethargic canines their repose. _____

9. Everything is dispassionate in infatuation and carnage. _____

10. Gaze at the environment as a precedent to sudden, upward, muscular movement. _____

Exercise 4 Revise the following messages to eliminate shifts and set a consistent, business-like tone.

   Example: I'm sending back this piece of junk.
   Revised: I am returning this product.

1. What's happenin'? I caught a jet to drop in on you and found you had cut out. I'd like to get a line on that new company you slave over.

2. I understand you have a prodigious assortment of topical technical periodicals. Your appraisal would be sincerely appreciated. Give me a jingle.

3. I won't tolerate any shilly-shallying on this undertaking. Vacillation will only retard the resolution.

4. We would like to hang out while you put on the feedbag and deliberate our druthers.

5. What do you envision as our game plan? Let's get the prerequisite go-aheads, like yesterday.

Answers may vary.

GRAMMAR UNIT TWELVE
AVOIDING SEXISM

In traditional writing, authors often used the masculine pronoun (he, him) in titles and examples that were actually referring to both men and women.

> Each student must submit his homework on time.

Years ago, this was accepted and tolerated. Currently, however, with the rising number of women in the working world and increasing awareness of bias and sexism, it is no longer acceptable to use just the masculine forms.

Even though we may suspect that the people we refer to in writing are either predominantly male or predominantly female, we must make attempts to include both genders. There are several ways to avoid sexism in writing.

1. Use pronouns of both genders. The pronouns can be separated by a slash (he/she, s/he) or a conjunction (or, nor).

> A student must submit his or her homework on time.
> Everyone gets as much sleep as he/she needs.

This method could become wordy and awkward with repeated uses.

2. Use singular, nongender pronouns. Some pronouns are impersonal, such as *one* or *that person*.

> When someone works hard, that person will be rewarded.

Again, this method is useful only in short pieces of writing or in occasional references.

3. Use plural (nongender) forms. The best neutral references are plural pronouns. Plural pronouns do not indicate gender.

> When people work hard, they are rewarded.
> Students must submit their homework assignments on time.

Remember that when subjects are changed from singular (he or she) to plural (they), sometimes the verb must be changed. "He or she *is* rewarded," or "They *are* rewarded."

4. Rephrase to avoid pronoun reference. Another way to eliminate the problem of pronouns is to reword the sentence into the passive voice so that no pronouns are needed. Remember that the passive voice weakens the statement since there is no "do-er."

> Hard work is rewarded.
> Homework assignments must be submitted on time.

5. Use "you" statements. Write directions, steps, or instructions using the "inferred you." Use only present-tense verbs.

> Work hard and be rewarded.
> Turn in homework assignments on time.

6. Use nonsexist terms, titles. The following sex-specific terms and titles are being revised due to increased sensitivity to nonsexist language.

Sexist	*Nonsexist*
chairman	chair, chairperson
committeeman	committee member
postman	postal carrier
policeman	police officer
Dear Sir	Dear Sir or Madam, Dear Personnel Manager
his	his or her, his/her (or change to plural)
he	he or she, s/he (or change to plural)
man, mankind	person, humanity, people, the average person
manpower	workers, employees, staff, personnel

Exercise 1 Edit the following passage for sexism. Do not change references to specific people.

Thomas Alva Edison was probably the greatest inventor in the history of mankind. After a mere three months of public school education, during a period when a schoolteacher considered herself aloof from questioning boys, "Al's" mother removed him from school and began teaching him at home. He asked an endless number of "why's" and received his first chemistry book at the age of nine. Soon he began to learn so fast that his mother could no longer teach him. Eventually, he was to patent 1093 inventions. Edison always tried to develop devices that would be useful to man, such as the electric light and the phonograph. He also improved the inventions of other men. These included the telephone, the typewriter, the motion picture, the electric generator, and electric-powered trains. He is known for his "brute force" method of solving problems: he would try endless methods, materials, and experiments until he found the answer, often working 18 hours a day in his lab. Edison believed that if a scientist was a genius, he would display "1 percent inspiration and 99 percent perspiration."

GRAMMAR UNIT THIRTEEN

TRANSITION WORDS

Writing a unified report begins with careful outlining and continues with writing headings and subheadings in the report. Because these methods are so essential, no other method can compensate for their loss. However, other methods can supplement them, and the easiest one is using **transition words** as you move from one idea or example to the next. The following charts show words that are commonly used to identify sequence of ideas, examples, contrast, and conclusions.

I. Sequence words indicate not only number order but emphasis or priority.

first	after that	in addition	especially
first of all	next	furthermore	particularly
second	later	last	moreover
third	then	finally	also, too

HINT

When listing reasons or cases, writers usually save the best for last. Begin with the weakest or least significant idea, and work up to the strongest or most important.

The example below shows one use of sequence words.

I follow the same steps when I write. First, I research my topic. Then, I write an outline. After that I start writing. The final and most important step is proofreading and editing. This last step, although it takes less time than the others, is the one I dread most of all.

II. Examples need to be clearly identified to prevent confusion and misunderstanding. When you are moving from an abstract theory or principle to a concrete application, provide immediate, obvious signals to your reader.

> for instance   for example
> that is   such as
> let us say   in the case of

Stress can have many side effects. For instance, a supervisor feeling pressure to complete a difficult project may become irritable, develop an ulcer, or possibly even succumb to illnesses such as a cold, flu, or more serious disease.

III. Contrast transitions show differences and similarities.

> however   nevertheless   moreover   still
> yet   instead   otherwise   even though
> on the one/other hand

The pressures of work and school were affecting my health. Nevertheless, I continued to keep up the pace.

IV. Conclusions will often identify the end of a section or report.

> after all   in conclusion   consequently   finally
> anyway   in fact   therefore   in summary
> at least   in short   thus
> at any rate   in other words   hence

The problems were not quickly resolved. Consequently, the project missed its deadline.

Exercise 1 Arrange the following transition words into the four categories listed below.

> then   for instance   therefore   thus
> nevertheless   however   finally   such as
> consequently   instead   particularly   for example
> let us say   in addition   in short   otherwise

Sequence	Examples	Contrast	Conclusion

Exercise 2 Choose logical transition words for the following passages. If more than one answer could be used, notice the difference in meanings, choose one, and write the others in the margin.

1. The vacation provided some much-needed and long-overdue relaxation. _____, after one week back at work, I felt like I had never been away.

2. The vacation provided some much-needed and long-overdue relaxation. _____, when I returned to work, I felt enthusiastic and energetic again.

3. The vacation provided some much-needed and long-overdue relaxation. _____ when I returned to work, I felt several changes. _____, I didn't drink as much coffee. _____, I listened better to what my colleagues were saying. And _____, I regained my sense of humor and found I could laugh at myself _____ getting defensive and hostile.

Exercise 3 Use sequence words to describe, in one paragraph, an activity you perform often from the following topics.

- Study for a test
- Perform your job
- Complete a lab

PART VI

MECHANICS UNITS

1. *Commas*
2. *Apostrophes*
3. *Quotations*
4. *Other Punctuation Marks*
5. *Abbreviations and Acronyms*
6. *Capital Letters*

MECHANICS UNIT ONE

COMMAS

Commas are punctuation marks (,) that are used for many different purposes. The rules for using commas are concrete, but since there are so many, remembering them requires some practice. This unit will review the basic rules for commas and give you practice in correct placement of commas. Knowing the rules will help make comma decisions easier in writing. Commas are used with the following:

1. Series of elements
2. Adjective pairs
3. Interrupters
4. Introductory information
5. Quotes
6. Compound and complex sentences
7. Miscellaneous uses

---------------- **NOTE** ----------------

Commas are always placed directly after a word.
Leave one space after a comma before beginning the next word.

COMMAS IN A SERIES

Two similar types of words written together are called a **pair.** They are separated by a conjunction such as *and* or *or*. A pair can consist of words or phrases.

> gray *or* blue *(pair of adjectives)*
> into the building *and* through the door *(pair of prepositional phrases)*

No commas are placed between pairs.

275

A **series** is three or more like words or phrases. Items in a series are separated by commas, with a conjunction before the last item in the series.

> gray, blue, *or* brown

We could add conjunctions between the items, but it would be wordy.

> gray *or* blue *or* brown

Instead, we eliminate all but the last conjunction and add commas between each item. Sentences can contain series of words and phrases.

> He bought new shirts, ties, shoes, *and* socks.
> He walked to the building, through the door, *and* inside the room.
> He introduced himself, shook my hand, *and* sat down in the chair.

Note Some people prefer not to put a comma before the conjunction in a series. Most technical writers do put a comma before the conjunction because it provides a visual cue, like green light turning to yellow, that the series is ending. Decide how you will handle it, and be consistent.

Exercise 1 Place commas in any series below. In this exercise, place a comma before the conjunction in the series.

1. On his first second and third interviews, he wore the same suit with different shirts and ties.
2. Margaret was interested in finding a navy-blue suit a blue blouse and black shoes.
3. The clerk asked Jack for his size and color preference.
4. To determine a company's dress code, you must read the employee manuals ask questions and watch the people around you.
5. The book recommended a gray suit white shirt black shoes and burgundy tie.
6. A "power suit" will make you feel confident attractive and proud of your appearance.
7. People looking for a new job a promotion or a salary raise must choose their wardrobes carefully.
8. Dave's weight and height limited his choice of styles.
9. Finding the right suit took time energy money and patience.
10. The personnel manager commented favorably on Roger's résumé interview and appearance.

COMMAS BETWEEN ADJECTIVE PAIRS

When two or more adjectives are used together, sometimes a comma is placed between them and sometimes not. Read the following examples.

1. a dark, simple suit
2. a dark business suit

In the first phrase, there is a comma between the two adjectives. Either adjective could modify the noun separately (a dark suit, a simple suit). The phrase could be written, "a dark and simple suit." In fact, the two adjectives could be reversed, "a simple, dark suit," and still make sense. In this case, a comma is placed between the two modifiers.

In the second phrase, dark is modifying a "business suit." Saying "a dark and business suit" would sound ridiculous, and so would reversing the two adjectives, "a business dark suit." A comma is not placed between these modifiers.

RULE OF THUMB

If the adjectives can be reversed and still make sense, put a comma between them.

Notice the difference caused by the comma in the two sentences below.

She wore a soft blue dress.
She wore a soft, blue dress.

The comma in the second sentence tells us that the dress was soft (in texture) and blue. In the first sentence, the blue color is a soft (pastel) shade.

Exercise 2 Place commas between modifiers if they are needed.

1. A handshake is a common business gesture.
2. A firm straight handshake puts a meeting off to a good start.
3. If you receive a brief abrupt handshake in return, the meeting is already in trouble.
4. Originally, a mutual clasp of forearms meant there were no small hidden weapons under the sleeve.
5. At an important business meeting, if everyone gets a handshake but you, be ready for bad news.
6. Years ago, fathers taught only their sons to shake hands because it was considered a required masculine gesture.
7. Today, women in business sometimes extend their hands first to eliminate any hesitation an unsure inexperienced man might have.
8. A vertical robust handshake signals you are ready to do business.
9. A frail limp handshake indicates lack of confidence or enthusiasm.
10. An overly aggressive bonecrushing handshake is a sign of fear, resentment, or extreme competitiveness.

COMMAS AROUND INTERRUPTERS

Commas are placed around words that interrupt the flow of the main sentence, particularly if the interrupting words are between the subject and the verb of the main sentence. You can usually "hear" a pause, even if the words are read silently. Words, phrases, and clauses can interrupt the main sentence. Test to see if the words are essential or interrupting by removing them from the sentence. If the sentence still makes sense, then the words are interrupters, and they should be set off with commas.

Words may interrupt the meaning of a sentence. Often they define or rename a noun. Information that is added because it is helpful, but not essential, is set off with commas.

NASA, the space agency, has promised free space shuttle launches.

In the example above, "the space agency" is written to rename NASA. Notice the difference made by commas in the examples below.

My brother Richard attended the launch.
My brother, Richard, attended the launch.

The first sentence means that Richard, not another brother, attended the launch—the name is essential. The second sentence means that there is only one brother, his name is Richard, and he attended the launch—the name is not essential, therefore it is interrupting.

Phrases may be written between the subject and verb of a sentence. Commas are placed before and after phrases that interrupt the meaning of the sentence.

Multiwire boards, unlike other boards, have a flexible design.
(interrupting phrase)
The design flexibility of multiwire boards creates an added benefit.
(essential phrase)

Clauses may be written within the main clause of a sentence.

The technicians, who are about to graduate, are in the lab.

In the sentence above, the main clause is "The technicians are in the lab." The dependent clause "who are about the graduate" is written within the main clause. Determine the difference between the two sentences below.

1. The technicians who are about to graduate are in the lab.
2. The technicians, who are about to graduate, are in the lab.

In sentence 1, the dependent clause tells where you can find the graduating technicians. The clause is necessary for the meaning of the sentence, so no commas are used.

In sentence 2, the dependent clause states that the students in the lab are also those who are about to graduate. As you read the sentence, you can hear a pause at the commas. The dependent clause is interrupting the main clause with extra (nonessential) information, so commas are placed before and after the dependent clause.

Another type of interrupter is the name of the person or group who is reading (or hearing) the message. We set off the person or group being spoken to with commas.

Students, go to the lab.
Start the experiment, Jack, and record the data.

―――――――――― **CAUTION** ――――――――――

Remember that commas around interrupters are used in pairs. If you use one comma at the beginning of a word, phrase, or clause, use a second comma at the end.

Wrong: The technicians who are about to graduate, are in the lab.

Wrong: The technicians, who are about to graduate are in the lab.
Correct: The people, who were talking about space, are outside.

Exercise 3 Add commas around interrupting words, phrases, or clauses.

1. The manager of the company Mr. Mendez issued the new policy.
2. The workers many of whom were new employees seemed to accept the change.
3. Soon more policies unrelated to the first occasion were changed.
4. The manager knowing that the workers weren't fixed in old ways took the opportunity to make changes.
5. That situation workers accepting change without resistance was unusual.
6. Dealing with resistance something new managers are unprepared for takes some experience.
7. I want you Mr. Riley to learn from this experience.
8. The president meaning no harm asked the manager to defend the changes.
9. The board of directors too were interested in the manager's methods.
10. Mr. Mendez however was eager to discuss the positive experience.

Exercise 4 Each of the following complex sentences has a dependent clause between the subject and verb of the main clause. Some are essential; some are interrupting. If the sentence is correctly punctuated, write CORRECT. If not, add commas where they belong. Be prepared to justify your answer.

1. Cellular telephone service which allows phones in motion to communicate with each other via radio is barely two years old.
2. Twelve companies that are competing for the first operating license are each determined to get a head start in selling to the new market.
3. The new service which will be called the mobile satellite system (MSS) will reach every part of the continental United States.
4. The first MSS operator that is licensed by the FCC will get free space shuttle launches.
5. NASA which promised the free shuttle launches will receive free satellite time in return.
6. The president of one communications company that is competing for the license claims the market is waiting for the technology.
7. Currently, 11 percent of the United States which is barricaded by mountains or a poor economy has no telephone service at all.
8. With MSS, though, even an Idaho lumberjack that uses a cellular terminal can receive and transmit messages.
9. The messages even though they come from the wilderness will travel via satellite to the telephone network used by the rest of the country.
10. Industry and emergency services which include nationwide paging, data collection, and position locating will become possible as a benefit of MSS.

COMMAS AFTER INTRODUCTORY INFORMATION

Words, phrases, and clauses are sometimes written to introduce the main sentence. Put commas after introductory information, as in the examples below.

> Students, gather up your equipment.
> Before long, the room was empty.
> Although they weren't finished, the students left.

--- **NOTE** ---

If the introductory words are short, a comma does not have to be added. (In the following exercises, however, add the commas.)

Exercise 5 Add commas after introductory information.

1. Over the past several years the interconnect industry has changed.
2. Most specifically the printed circuit board fabricators have been informed of advanced packaging in semiconductors.
3. Resulting from these changes PC boards must change to accommodate semiconductors in areas of electrical speed and rise times.
4. Although these comments have been found to be true PC fabricators have received little assistance in keeping up with the changes.
5. More to the point PC fabricators need to find practical and realistic means to control their products.
6. As is normally the case in this type of situation the answer lies in a compromise with both parties contributing their insight.
7. For example this situation is at the heart of PCB-controlled impedance.
8. Although this technology is many years old the only positive assurance of a "good" board is an "end of line" TDR test.
9. If you are unfamiliar with this test it is called the GO/NOGO.
10. Unfortunately the NOGOs go into the junk heap.

Exercise 6 Edit the following paragraph for commas. Look for interrupters and introductory information. You should find 15 missing commas.

Swathed in clouds and pelted by rain a small plane struggled for altitude as it approached the highest peak in Alabama Mt. Cheaha. In the late afternoon of November 19, 1984 the single-engined Cessna slammed into the trees. The pilot a 53-year-old man was trapped in the wreckage with severe injuries. His wife and child both unhurt started a fire and the emergency radio beacon. They were rescued the next morning. Their distress beacon as they were to find out later was picked up by a solar-powered Russian satellite spinning 600 miles overhead. The international satellite rescue system which was established during the years of detente as few years earlier has helped save hundreds of lives around the globe. The satellites listening for distress calls from emergency transmitters relay the automated calls for help to nearby ground stations. These satellites thought of as celestial Samaritans are above politics in more ways than one.

Check your corrected paragraph with your instructor.

COMMAS IN DIRECT QUOTES
Commas are used to separate a direct quote from the rest of the words in the sentence. Commas are placed directly after a word, before the beginning or ending quotation mark. For more information on quotes, see "Quotations."

> In its report, the committee stated, "Our recommendation is to widen our market."

Do not use commas to set off quoted material that is only part of a sentence.

> The committee suggested that we should "widen our market."

Exercise 7 Add commas to the direct quotes.

1. The critic stated "This foul-up is the worst I've seen."
2. "What's more" he said "new developments surface every day."
3. "But, despite numerous setbacks, we're making progress" he said.
4. "The latest protest" he added "comes from within our ranks."
5. "Many times, we are our own toughest critics" he mused.

COMMAS IN COMPOUND AND COMPLEX SENTENCES
These rules are covered more thoroughly in the grammar units covering compound sentences and complex sentences. As a brief review, remember the following rules.

1. Use a comma and a coordinator to fuse two complete sentences into a compound sentence.

   There are no perfect insulators, for all insulators have leakage current.

   The common coordinators are AND, BUT, FOR, YET, OR, SO, and NOR.

2. Use a comma after a dependent clause that introduces the main clause in complex sentences.

   Because all insulators have leakage current, there are no perfect insulators.

   Do not use a comma if the main clause comes first.

   There are no perfect insulators because all insulators have leakage current.

   Some common signal words for dependent clauses are IF, WHEN, BECAUSE, SINCE, ALTHOUGH, WHO, WHICH, and THAT.

Exercise 8 Add commas where necessary.

1. Specific codes provide error-inherent detection and you will discuss several of them.
2. When you check for or generate proper parity in a given code word a basic principle is used.

3. The sum (disregarding carries) of an even number of 1's is always 0 and the sum of an odd number of 1's is always 1.
4. To determine if a given code word is even or odd parity all of the bits in that code word are summed.
5. The sum of two bits can be generated by an exclusive-OR gate and the sum of three bits can be formed by two exclusive-OR gates.

MISCELLANEOUS USES OF COMMAS

Commas are used in ordinary ways, such as in dates, addresses, openings and closings of letters, numbers, and contrasted material. Read the following examples.

Dates: He was hired on August 15, 1985, in Seattle.
I read the March 1987 issue on Friday, May 5, 1987.
OR: I read the March 1987 issue on Friday, 5 May 1987.

---------- **NOTE** ----------

The commas are not required when the day of the month is missing, or when the date is written in international style (day, month, year).

Addresses: Miami, Florida, is the home of the Dolphins.
The address is 4217 Lincoln Drive, Minneapolis, Minnesota 55455.

---------- **NOTE** ----------

Commas are not used between a state and the zip code.

Write a letter to Mr. Thomas Samuels, Jr., El Paso, Texas.

Letter openings: Dear Mr. Samuels,   Dear Emily,

---------- **NOTE** ----------

In formal letters, a colon is used after the greeting.
Dear Mr. Samuels: Dear Ms. Matthys:

Letter closings: Yours truly,   Sincerely,
Numbers: We received over 1,000 orders.
Contrasted material: The meter read 100 mA, not 1,000 mA.

Exercise 9 Add commas where necessary.

1. I applied for the job on May 19 1985.
2. I received a note that my résumé was being forwarded to Dr. Sherrit Ph.D. College Recruiter San Jose California.
3. She wrote back in November 1985 to set an interview in the Palo Alto California plant for January 15 1986.
4. Then she changed the date to February 1 not January 15.
5. She explained that over 1500 college students had applied for the 50 available positions and signed the letter "Sincerely Ms. Sherrit."
6. The plane for San Jose California left on January 31 1986 at 10:15 a.m.
7. The hotel was at 33245 El Camino Street San Jose.

8. The interview was in Suite 616 1899 S. San Miguel Street Palo Alto California.
9. I listed by home address as Marshall Minnesota 56258.
10. The company had over 60000 employees working in 1250 plants.

A FINAL WORD ABOUT COMMAS

Many writers, especially beginning writers, recall being told to use a comma wherever they pause. Sometimes this generalization is useful, but more often it adds to the confusion and uncertainty of when to use commas, and how many commas to use. It can also lead people to believe that there are no firm rules, that only breathing patterns determine comma placement. The truth is that you can pause wherever you see a comma, and that is about as far as breathing is involved in the matter.

Although it is true that exceptional cases can break the rules, and that professional authors seem to honor some rules and ignore others, commas are normally placed according to standard rules established to help the reader. Using or not using a comma can change meaning, as examples have shown. Use commas carefully, as tools to make your message clear.

Exercise 10 Edit the following sentences for commas. Cross out or add commas as needed.

1. Computers are habit forming; they can draft reports, tally figures query information services and, order airline tickets.
2. But as the habit grows access to computers in the home and office is no longer, enough.
3. Today's computer users want to be able to draft memos on a plane, or to dial into the office computer from a phone booth.
4. In short, they need a computer, to take along wherever they go.
5. The first "portable" computers, were not portable at all but they were "luggable."
6. By far the heaviest, and bulkiest part of the package, was the cathode-ray tube (CRT), for the information display.
7. Three main alternatives, to the cumbersome CRT display, are the liquid-crystal, electroluminescent, and plasma, displays.
8. In a liquid-crystal display the surrounding lighting conditions, are critical to readability.
9. Electroluminescent displays (ELDs) consume more power, and cost much more, because higher-voltage drive circuits are required.
10. Plasma displays use a gas to convert gas, into energy and they consume the same amount of power as electroluminescent displays.

MECHANICS UNIT TWO
APOSTROPHES

Apostrophes (') have two common uses: contractions and possessives.

CONTRACTIONS
Contractions are combinations of two words, similar to some abbreviations. However, rather than using the first letters, we combine the two words to form one new word. Usually we keep the beginning or all of the first word and combine it with the last part of the second word.

should not	shouldn't
I would	I'd
she has	she's
she is	she's

Notice that the apostrophe is added exactly where the letter or letters are dropped. The only common exception is "won't," the contraction of "will not."

Contractions are used more often in informal writing, where the two full words would sound awkward and formal. In formal reports, use contractions as seldom as possible. If you do use them, be sure that the words are contracted correctly.

─────────── **NOTE** ───────────

Apostrophes are used to indicate any intentionally dropped letters or numbers:
   the class of '87 (1987)
   we're comin' (coming)
   o'scope (oscilloscope)

Exercise 1 Write the contractions for the following pairs of words.

1. he would _____
2. is not _____
3. I have _____
4. it is _____
5. they are _____
6. are not _____
7. they have _____
8. cannot _____
9. you are _____
10. should not _____

Exercise 2 Use each contraction from Exercise 1 in a sentence. Proofread each sentence to make sure that you can correctly substitute the two uncontracted words.

1. _____
2. _____
3. _____
4. _____
5. _____
6. _____
7. _____
8. _____
9. _____
10. _____

POSSESSIVES

Possessive apostrophes indicate ownership. The apostrophe changes a noun to an adjective. Read the following sentence.

> Sam drove his father's car.

We know that the car Sam drove belonged to his father. We have turned "father" (a noun) into an adjective to describe a certain car. When we turn a noun into a possessive adjective, we add an *apostrophe S* to the end of the noun. Usually the sentences could be reworded stating the ownership as a prepositional phrase.

> Sam drove the car of his father.

If you are not sure of whether or not to add an 'S to the end of a word, try rewording the sentence, putting the name of the "owner" in a prepositional phrase.

Exercise 3 Reword the following possessives stating the object owned, followed by a prepositional phrase.

Example: neighbor's yard
Rewritten: yard of the neighbor

1. week's salary _____
2. resistor's value _____
3. student's grades _____
4. runner's time _____
5. lab's hours _____

Apostrophes

Exercise 4 Add apostrophes if necessary.

1. The first tests results were recorded.
2. My new voltmeters readings seemed slightly off.
3. The tests were returned after only one weeks time.
4. We used Faradays law of induction.
5. The number of farads indicated the amount of capacitance.
6. The number of capacitors in that circuits schematic was surprising.
7. The amount of capacitance is determined by each capacitors design.
8. When capacitors are in parallel, each capacitors voltage is equal.
9. The electrostatic fields energy depends on capacitance and voltage.
10. When discharging, capacitors return energy to their circuits.

Plural Possessives Notice that in each case above, the "owner" was singular. When an owner is plural, we add the apostrophe after the plural form of the word.

> the workers' schedules (the schedules of the workers)
> the children's room (the room of the children)

If the plural owner already has an "S" at the end of the word, we don't have to add an extra S after the apostrophe, although some writers do.

> the cities' problems (the problems of the cities)
> the ladies' club (the club of the ladies)
> Chris' turn (the turn of Chris)

Exercise 5 Using the phrases from Exercise 4, rewrite each possessive phrase with a PLURAL owner.

> *Example:*   the neighbor's yard
> *Rewritten:* the neighbors' yards

1. week's salary _____
2. resistor's values _____
3. student's grades _____
4. runner's time _____
5. lab's hours _____

Some possessives do not need apostrophes, and these are the **possessive pronouns** such as *hers, his, yours, its, ours,* and *theirs*. These pronouns are usually used as adjectives, but they do not have an apostrophe.

> *Examples:* The car is *hers*.
> *His* job is difficult.
> Hold the hammer by *its* handle.
>   [do not confuse "its" for the contraction "it's" (it is)]
> The mistake was *theirs*.

Special Cases

1. For compound words or word groups, add the 'S only to the last word.

    Example: someone else's idea
    Chief Executive's decision
    father-in-law's consent

2. Add an 'S to each individual owner.

    Example: teacher's and students' experiences
    workers' and managers' responses

3. To show joint ownership, you can either add an 'S to both names or only to the last name.

    Example: Mr. Nguyen and Mr. Lee's reaction
    (the same reaction)
    or: Mr. Nguyen's and Mr. Lee's reaction
    (the same reaction)

Exercise 6 Add the possessive apostrophes where needed. There are 11 apostrophes needed.

Many handicapped people have learned how to turn their disabilities into victories. One such person is my friend, Phil, a junior in an electronics engineering technology program. Phils eyes are extremely weak, leaving him nearly blind. He is not able to see objects more than 2 inches away. He cannot read a typewriters keys or a tests words without bumping his nose into them. Somehow he can push his calculators keys while looking at the readout. He listens to his teachers lectures, and later he reads a classmates notes written with a dark pencil. His textbooks are recorded on cassettes, and his two lab partners assist him in all his labs safety procedures. Phils GPA is over 3.00. Last term, he was placed on the Deans List. We all admire Phils and many other handicapped students strength and determination to overcome their disabilities.

MECHANICS UNIT THREE

QUOTATIONS

Many technical writers find it necessary to quote other authors or speakers. A **quotation,** or **quote,** is information repeated or reproduced from another source. Quotes can add interest, precision, and credibility to technical material. Quoted material includes information repeated in the exact words of the original speaker or writer (**direct quotes**) or slightly reworded or summarized accounts of the original version (**indirect quotes**). Both types of quotes are cited in a footnote, endnote, or parenthetical note (refer to Appendix 3), but only direct quotes have special punctuation. Failing to cite the source of either a direct or indirect quote is considered **plagiarism,** a serious writing offense. Avoid plagiarizing by using quotations marks (in direct quotes) and citing the source (in direct and indirect quotes).

DIRECT QUOTES

The exact words of a speaker or writer are surrounded by **quotation marks** ('').

"Robots are just a combination of hydraulics, mechanics, electronics, and computers."

Sometimes the speaker is included in the sentence. The quotation marks still surround the quote, but the speaker is set off by a comma.

"Technical specialists are very expensive," notes Nancy Johns.
"Technical specialists," notes Nancy Johns, "are very expensive."

RULES FOR WRITING DIRECT QUOTES

1. Begin every quoted sentence with a capital letter even if the first word of the quote is not the first word of the sentence.

One assembly worker said, "That thing?"

Quotations 289

Do not begin quoted words or phrases with a capital letter unless other reasons require a capital letter.

> A "dedicated robot" is one that has been permanently programmed to perform certain tasks.
>
> "Dedicated robots" are those which have been permanently programmed to perform certain tasks.

2. Place the first set of quotation marks immediately next to the first word of the quote (no spaces).

3. Put the closing set of quotation marks at the end of the quote and after any other punctuation. If the quote ends the sentence, put the final quotation mark outside the period or question mark.

> He asked, "How did you know?"
>
> *Exception:* If the whole sentence is a question, but the quote is not a question, place the question mark after the closing quotation mark.
>
> Did the manager say, "Take the day off"?

If the quote is followed by other information, end the quote with a comma and the second set of quotation marks.

> "The manager just left," the worker observed.
>
> *Exception:* If the quote ends in a question mark or exclamation mark, keep it.
>
> "What did he say?" asked the worker.

If the speaker's name interrupts the quote, two pairs of quotation marks must be used.

> "I heard him say," responded my partner, "that we have tomorrow off."

4. If you are including a parenthetical note (see Appendix 3) at the end of a quoted sentence, place the final quotation mark before the first parenthesis.

> "Robots aren't replacing entire shifts of workers" (Fey, p.49).

5. Use single quotation marks (') around a quote within a quote.

> Hoska reports, "I was called into plants by management and told, 'Find us a place to use a robot.'"

6. Do not use a semicolon before a direct quote.

7. Place the closing set of quotation marks at the end of the entire quote if it consists of more than one sentence.

8. Use quotes sparingly or not at all. Instead, try to write ideas in your own words.

RULE OF THUMB

Do not use more than one direct quote per page.
Do not quote more than one paragraph at a time.

Quotations

NOTE

Quotes that are formal or long are sometimes introduced with a colon.

> The manual includes the following caution: "Do not connect the power cable to the instrument before verifying that the intended source matches the ac line configuration of the instrument."

Exercise 1 Add quotation marks and commas where necessary. Keep each sentence a direct quote.

1. It took a human mind to invent the wheel and build a computer said my professor.
2. He added now people can make a machine with a mind of its own.
3. He asked us what can robots do?
4. I answered I think they can cut, weld, paint, and lift things.
5. That's right he said but someday soon they will do much more.
6. Someday soon they may clean up toxic wastes, disarm bombs, or explore the ocean floor he suggested.
7. Did you say disarm bombs? I asked excitedly.
8. Yes he replied and much more.
9. I then asked what is it that makes a robot a robot?
10. Most of us agree said my professor that a robot must be two things. It must be mobile, and it must be programmable.

Exercise 2 Write five direct quotations of your own using correct placement of quotation marks, commas, and ending punctuation.

1. _____
   _____
2. _____
   _____
3. _____
   _____
4. _____
   _____
5. _____
   _____

INDIRECT QUOTATIONS

Indirect quotes state the general ideas of another speaker or author, but not the exact words. Read the following indirect quotes.

> Nancy Johns noted that technical specialists are very expensive.
> Troubleshooter Bob Adams reports that when a crew heard that a Puma robot was going to be installed, people feared for their jobs.

Indirect quotes may be close to a direct quotation except for an added "that," or they may be condensed accounts of a longer direct quote. Note that no special punctuation is used for indirect quotes, which makes them easier to write. Remember, however, that indirect quotes still need to be cited in research papers either by a footnote, endnote, or parenthetical note.

PLAGIARISM

Many technical reports are based on the research of other people. When writing about a new or difficult subject, some writers are unable to reword their source. Most writers understand the obligation to cite the source of a direct quote. What is less understood, however, is how to handle a passage that has been slightly reworded.

Passages that have essentially been copied with slight word or structural changes are still considered indirect quotes and should be cited.

A passage is an indirect quote if the following situations apply:

1. The order of ideas is the same. No additional information is added.
2. The words or sentence structure is changed, but the ideas are the same.

Read the following direct quote and a paraphrased, indirect quote. Both should be cited as quotations. The passage is taken from "Machine Vision," by Nello Zuech, *Robotics Today,* April 1986, p. 35.

> *Direct quote:* "The 1986 market for machine vision should exceed $200 million, a 33% growth over 1985. If one considers products with dedicated performance envelopes, such as photomask inspection systems used in microelectronic manufacturing or off-line dimensional measuring equipment, the market will easily exceed $300 million" (Zuech, p. 35).
>
> *Indirect quote:* Machine vision markets in 1986 should grow by 33% over the 1985 market, reaching an excess of $200 million. Dedicated performance envelopes such as photomask inspection systems may bring the market to $300 million (Zuech, p. 35).

Exercise 3 Rewrite the direct quotes from Exercise 1 as indirect quotes. Remember to make the indirect quotes sound natural and smooth. The first one is done for you.

1. "It took a human mind to invent the wheel and build a computer," said my professor.

    My professor said that it took a human mind to invent the wheel and build a computer.

2. _____

3. _____

4. _____

5. _____

6. _____

7. _____

Quotations

8. _____
9. _____
10. _____

Check your answers with your instructor.

OTHER USES OF QUOTE MARKS

1. Quote marks are used to set off titles of articles, stories, songs, or poems—italicize or underline titles of books or magazines.

   The article "Shuttle Communications" is in the summer issue of *Hands-On Electronics*.

2. Quote marks are used to show words with a special or unusual meaning.

   Abraham "Honest Abe" Lincoln
   Ice (especially "dry ice") can cause tissue damage if applied directly to a burn.

———— NOTE ————

Do not use quotes around common nicknames.

Once a word has been highlighted with quotes, continued usage of that word does not require quotes.

Exercise 4 Add quote marks where necessary.

1. The Technically Speaking column appears in most issues of *IEEE Spectrum*.
2. One issue talked about nouns that are being used as verbs, such as keyboard and messenger.
3. Add automated reasoning and smart machine to the list of items coming into frequent use from the field of artificial intelligence.
4. One reader asked if anyone could verify the following account:
   At one time the British Labor Party, in its annual convention, came within a few votes of passing a resolution that stated No one shall receive less than the average wage.
5. I then looked up Great Britain in the encyclopedia.

MECHANICS UNIT FOUR

OTHER PUNCTUATION MARKS

Punctuation marks other than commas, apostrophes, and quotation marks will be reviewed in this unit. This unit includes the following:

- Colon
- Semicolon
- Parenthesis
- Dash
- Bracket
- Slash
- Hyphen

COLON

1. The colon (:) is used to introduce lists of items following a complete sentence.

> Four types of communication are the following: listening, reading, speaking, and writing.

--- **NOTE** ---

The colon is never used following a verb. After a verb, simply state the list as a series of items.

The four types of communication are listening, reading, speaking, and writing.

2. Colons are used to express ratios of numbers. We sometimes read these colons as the word "to."

> His odds were 2:1.

3. Colons are used in formal business letters in place of the comma following the greeting.

    Dear Sir or Madame: Dear President Reagan:

4. Colons are used according to conventions when indicating times, biblical passages, and subtitles.

    9:15 p.m.   12:00 noon   Mark 2:1–15
Modern Electronics: Basics, Devices, and Applications

SEMICOLON

1. Semicolons (;) are most frequently used to combine two related, complete sentences into one compound sentence. Think of a semicolon as a link between the period and the comma—both are included in the symbol.

A semicolon is placed at the end of the first sentence, followed by one space. The first word of the second sentence is not capitalized (unless it needs to be capitalized for another reason).

    His new job was satisfying; he felt productive and challenged.

2. Semicolons are sometimes used between sentences linked by words such as "therefore," "however," and "moreover."

    He worked at that company for six years; however, he held a variety of positions during that time.

3. Semicolons are used instead of commas between a series of items that contain commas within the items.

    The courses in the first three terms are challenging: Digital I, Electronics I, and Technical Math I; Digital II, Electronic Devices, and Technical Math II; and Electronic Communications, Intro to Microprocessors, and Industrial Control Systems.

Exercise 1   Add colons or semicolons where needed. Do not add any other type of punctuation.

1. The United States operates five satellite tracking stations Goldstone, CA Kauai, HI Merrit Island, FL Fairbanks, AK and Rosman, NC.
2. The Challenger was launched before 8 00 a.m. in April 1983 with it was launched the first Tracking and Data Relay Satellite (TDRS).
3. The orbiting satellites have two sets of antennas the S-band and the Kμ-band.
4. The S-band antennas are located on the outside surface of the shuttle Kμ-band communications use a steerable dish antenna mounted in the cargo bay.
5. Before TDRS, the ratio of time in and out of contact with a shuttle was 1 4.
6. With TDRS, nearly constant global coverage is possible therefore, this tracking system can replace more than half of those on the ground.
7. Mission Control at Houston is the central point for shuttle flights here the communications flexibility is greatest.

8. The interface systems at Mission Control use computers for two reasons to format, compress, and route outgoing information and to reformat, decode, and forward incoming data.
9. Other communications systems are connected to NASA the Kennedy Space Center, Cape Canaveral, FL and Edward AFB, CA.
10. The Defense Department provides logistics, special studies, and search-and-rescue support these are performed by the Army, Air Force, Navy, and Coast Guard personnel and equipment.

PARENTHESES

1. **Parentheses** (plural form of "parenthesis") are used to set off interrupting information. Because parentheses are used less frequently than commas, parentheses provide more of a separation than commas. In fact, the expression "parenthetical remark," which refers to related but nonessential information, comes from the function of parentheses.

Parenthetical remarks interrupt the flow of the sentence, and they usually can be "heard" as a pause even when read silently. They enclose extra information, comments, or facts.

> Some people use big words to impress others (or so they think) at the cost of being clear.
>
> A mass announcement (high efficiency) dealing with a sensitive issue that could better be dealt with through an individual discussion (low efficiency) can result in low effectiveness.

2. Parentheses are also used when the added material is too long to be inserted with commas, such as complete sentences or long phrases, particularly those that include commas.

> We increase efficiency by cutting the message's cost (for example, running off multiple copies rather than typing original letters) or reaching more people (broadcasting a message to all workers in an office rather than selected audiences).

3. Parentheses can be used to enclose numbers or letters that enumerate ideas in a sentence. They can be used singly or in pairs for this purpose.

> Failing to close the feedback loop can take several forms: (1) not listening to our own messages, and (2) not listening to our receiver's feedback.

RULES

Remember to follow the punctuation rules when using a set of parentheses:
A. Place no punctuation before the first parenthesis.
B. Delay the punctuation that would have come before the first parenthesis until after the second parenthesis.
C. Follow standard punctuation and capitalization rules within the parenthetical remark.

DASH

1. The **dash** is the most abrupt separator of interrupting information. It is used in places where commas or parentheses could also be used, but it adds the most emphasis.

> Ask for feedback to see how well you're doing—kids are great teachers.
>
> Studies have shown that as much as 78 percent of meaning is transmitted nonverbally—that is without words.

2. A dash is occasionally used in place of a colon to introduce a list.

> We have a variety of communication media available for our use—telephones, memos, letters, interviews, group meetings, and so forth.

3. A dash can also be used to "access" or highlight key ideas or bits of information.

> There are four essential functions of nonverbal messages:
> - to accentuate information conveyed verbally
> - to express like or dislike
> - to convey intensity of feeling
> - to contradict verbal messages

_____ **NOTE** _____

On the keyboard, a dash is made with two unspaced hyphens. In handwriting, make the dash longer than a hyphen.

BRACKET

Brackets are squared-off parentheses that are usually used within direct quotations. Brackets are used only in pairs.

1. Brackets could be used because you want to add extra or specific information or modify a quote, which is often necessary when the quoted sentence includes pronouns that were defined elsewhere.

> *Original:* "We will cover three major changes in our compensation plan."
>
> *Modified:* "[The board of dirctors] will cover three major changes in [Acme Incorporated's] compensation plan."

2. Use brackets around the Latin word *sic* (meaning "thus") after an author's error to show your reader that it was not your error.

> *Spelling error:* "Failing to close the feedback look [*sic*] can take several forms."

_____ **RULES** _____

The rules for punctuation around or within parentheses also apply to brackets. If your keyboard does not have a bracket symbol, draw it neatly with a pen.

ELLIPSIS

The **ellipsis** (consisting of three spaced dots) is used to mark words left out of direct quotes.

> *Original:* "The shuttle voice communications with the ground use the S-band 10-watt (2205.0 and 2250.0 MHz) transmitter."
>
> *Modified:* "The shuttle voice communications with the ground use the S-band 10-watt . . . transmitter."

SLASH

1. The **slash** is used to indicate choices or options. It is generally used to replace the word "or." Occasionally, it is used in ambiguous situations to allow for the possibility of singular and plural events (result/s) or male and female individuals (s/he). The slash should not be used too often for these purposes.

> Some courses are not taken for a grade, but just for a pass/fail credit.
>
> A manager will cause the death of his/her effectiveness by committing the seven deadly "sins" of management communication.

2. The slash is frequently used in dates consisting of numbers.

> The order was dated 11/20/87.

_____ **NOTE** _____

An enumerated date should be written and read as month/day/year.

3. In technical notation, the slash symbolizes division.

> $R = V/I \quad I = V/R$

HYPHEN

1. The **hyphen** (-) is usually used to join two words. Some "compound" words always have a hyphen between them. Dictionaries will include required hyphens.

> ampere-turn   high-pass
> hard-core   full-scale

Dictionaries sometimes disagree about hyphenated compound words, so be consistent in your own writing. Generally, follow these rules:

A. Hyphenate compound modifiers preceding a noun, but do not usually hyphenate those modifiers if they follow a noun.

> He was a well-dressed man.
> The man was well dressed.

Always hyphenate technical terms that require hyphenation:

> The pulse is positive-going.
> ohms-per-volt rating
> T-type low-pass filter

Other Punctuation Marks

   B. Do not hyphenate compound modifiers if one of the words ends in -LY.

   He was a professionally dressed man.

   C. Hyphenate all compound numbers between twenty-one and ninety-nine and some fractions.

   one-tenth   two-thirds
   one half   one quarter

2. Hyphens are sometimes used to divide a word at the end of a line. Remember that hyphens must be placed between syllables (use your dictionary to verify syllables), and one-syllable words cannot be divided. It is common practice not to begin a line with short syllables. Instead, write the entire word on the next line. Divide hyphenated words only at the hyphen. Avoid two consecutive lines ending in a hyphen.

   Wrong: The difficult and time-consum-
              ing lab gave me a terri-
              ble headache.
   Correct: The difficult and time-
              consuming lab gave me a
              terrible headache.

3. Hyphens, like dashes, can be used to highlight information.

   To simplify circuits for analysis, we can use many methods:
   - Superposition
   - Thevenin's theorem
   - Norton's theorem
   - Millman's theorem

Exercise 2 Correctly punctuate the following sentences by adding parentheses, dashes, brackets, hyphens, or slashes. Be prepared to defend your answer, especially for notation setting off parenthetical remarks.

1. The announcement stated, "I want all the machines that are broke sic sent back to the lab."
2. The test results will indicate GO NOGO.
3. His last job I haven't the slightest idea of his other jobs was in the production department.
4. I tried every two number combination between thirty five and sixty five.
5. We received the replacement on 4 20 87.
6. Mr. Harcort the manager and Mr. Leninson the job foreman arrived for the appointment.
7. A capacitor and an inductor can be connected in an inverted L configuration.
8. A band pass filter it will follow a typical band pass response curve allows a certain band of frequencies to pass.
9. The total ohmmeter resistance is 1.5 V 50 μA = 30 kΩ.
10. When the ohmmeter leads are open, the pointer is at full left scale, indicating infinite resistance open circuit.

MECHANICS UNIT FIVE

ABBREVIATIONS AND ACRONYMS

Shortened forms of words and phrases are referred to as **abbreviations** or **acronyms**. For all but the most common shortened forms, technical writers provide the expanded meaning of an abbreviation or acronym *the first time it is used*. Either the expanded meaning or the acronym may be placed in parentheses.

The circuit needed a *direct current (dc)* voltage source.
The computer has *64K (kilobytes)* of *ROM (read-only memory)*.

Many technical symbols and jargon are abbreviations and acronyms. They do not follow the standard rules for capitalization, since sometimes a small letter may stand for one word and a large letter may stand for another, such as *M (mega)* and *m (milli)*. Many common electronic symbols and abbreviations are listed in Appendix 1. Common abbreviations and acronyms are found in some dictionaries and textbooks. Be sure to use technical abbreviations, acronyms, and symbols properly.

Abbreviations are shortened forms of words or phrases. Each letter (as in CIA) or each unabbreviated word (as in titles) is pronounced. Abbreviations of titles such as "Mr." or "Jr." begin with a capital letter and end with a period. Pronounce the title as though the full word was written.

Other abbreviations, such as "TV," "ac," or "FBI," may stand for one or several words. They are pronounced either by the letters (if generally understood) or by the full words.

The abbreviations for states have undergone a recent change. Currently, we can use the two-letter postal code for states, such as GA (Georgia) or CA (California). Notice that both letters are capitalized and no period is added. Before this code was established, people abbreviated states by using the first few letters, capitalizing only the first letter, and ending with a period, as in Geo. (Georgia) or Calif. (California). Some states were abbreviated several ways (Cal., Calif., Ca.). Although some people still use the old method, the new postal code provides a uniform system.

Abbreviations and Acronyms

Exercise 1 Find the postal abbreviations for the following states (see Appendix 1).

Alabama _____ Alaska _____ Arizona _____ Arkansas _____

Write the postal abbreviation for your state. Write the postal abbreviations for three states in which you would like to live.

Your state _____   Others _____ _____ _____

Other abbreviations, such as days of the week and months, usually consist of the first syllable of the word, beginning with a capital letter, and ending with a period. If the word is short or has only one syllable, such as May or June, do not try to abbreviate it.

Exercise 2 Write the standard abbreviations for the days of the week (see Appendix 1).

Sunday _____   Monday _____   Tuesday _____   Wednesday _____
Thursday _____   Friday _____   Saturday _____

Exercise 3 Write the expanded form of the following abbreviations.

1. TV _____
2. mike _____
3. xtal _____
4. amp _____
5. op-amp _____
6. scope _____
7. mono _____
8. stereo _____
9. xmtr _____
10. D/A converter _____

Acronyms are words created by combining the first letter of most or all of a combination of words to make a pronounceable word. Many times these word groups are arranged and contrived to make the pronunciation easier, such as bit (binary digit) or laser.

Write the EXPANDED meaning of laser:

l_____ a_____ by s_____
e_____ of r_____

Even though acronyms are types of abbreviations, no periods are placed after the letters, and all the letters are usually capitalized. When acronyms such as laser become commonly known and used, they are written in small letters. Watch your textbooks and technical magazines for accepted forms.

Exercise 4 Write the expanded meaning of the following acronyms. Use a dictionary if necessary.

1. RAM _____
2. ANSI _____
3. DIP _____
4. bit _____
5. EPROM _____
6. MOSFET _____
7. NASA _____
8. sonar _____
9. scuba _____
10. radar _____

MECHANICS UNIT SIX
CAPITAL LETTERS

I

Several capitalization rules which are most often used in technical writing will be reviewed. Style guides are available in libraries and bookstores for specific or unusual situations.

A popular trend these days is using a company logo (emblem or symbol) of lower case (small) letters. This appears most frequently in "artistic" advertisements for products. Most writers, when referring to these same items, use standard capitalization rules regardless of the company's stylized logo.

    phase linear®
    I bought Phase Linear speakers for my car.

Rule 1 Captalize the first word of a sentence, the first word of a quote, and the pronoun "I."

    The elderly man and I didn't hear the last call.
    We heard the conductor say, "Last stop coming up."

Rule 2 Capitalize specific names of people, including initials and nicknames (although nicknames are rarely used in formal writing).

    Honest Abe Lincoln   John F. Kennedy
    Robert "Skip" Simons   B. F. Skinner

Rule 3 Capitalize names of countries, states, cities, and other geographic regions. Do not capitalize directional words.

    the Southwest   New York City
    turn southwest   European vacation

Rule 4 Capitalize names of races, religions, nationalities, and any word derived from the name of a country.

>Chinese Puerto Rican
>Episcopal Indian

--- **NOTES 1** ---

Some words derived from a country name are used so often that we sometimes forget to capitalize them.

>French fries Italian dressing
>Afro haircut English ivy

Most Americans use *white* and *black* to refer to races. These words do not have to be capitalized, though some writers prefer to capitalize them. Decide how you will treat them and be consistent.

Exercise 1 Change any incorrect lowercase letters to capital letters.

1. although robert grew up in florida, he now considers himself a georgian.
2. we learned that after the catholics were driven out of europe, they immigrated to northeastern regions in the united states such as new york, massachusetts, and connecticut.
3. my mother ordered italian pizza, french bread, and boston lettuce salad.
4. martin luther king said, "i have a dream."
5. there are many muslims and jews in new york city and many baptists in atlanta.
6. her sales territory covers the entire northeast.
7. my family and i moved from the north to miami, florida.
8. jazz is a popular form of music in new orleans.
9. thomas a. edison, credited for saying "there is no substitute for hard work" invented the thermionic diode.
10. silicon valley is a nickname for an industrialized area in northern california surrounding san jose.

Rule 5 Capitalize a person's title if a proper name follows it, but not when the title is used alone.

>Professor Henry Higgins former President Ford
>the professor a former president

--- **NOTE** ---

Abbreviated titles (Mr., Ms., Jr., Ph.D.) begin with a capital letter and end with a period.

Rule 6 Capitalize names of days and months, but not seasons.

>Thursday spring
>March fall

Rule 7 Capitalize names of languages and specific courses, but not general subject areas.

        electronics   mathematics
        Digital Electronics   Technical Mathematics
        English   Pascal

Rule 8 Capitalize a person's name which is part of a general theory or principle, but not general electronic or scientific terms, even if derived from a name.

        Ohm's law   5 ohms
        Einstein's theory of relativity

Exercise 2 Change any incorrect lowercase letters to capital letters.

1. freshman orientation was taught by a new professor.
2. the doctor spoke both english and spanish fluently.
3. all electronics labs will meet on tuesdays and thursdays during the fall term.
4. he studied the differences between thévenin's theorem and superposition.
5. our class will graduate from the electronics technician program in june.
6. the electron was discovered by french physicist jean baptiste perrin and british physicist sir joseph thompson.
7. we use ohm's law to find the relationship between resistance, voltage, and current.
8. joseph henry was an american physicist after whom the unit of inductance, the henry, is named.
9. my cousin was supposed to appear before judge carlson on tuesday, but the judge rescheduled the hearing.
10. my professor earned a ph.d. last fall.

Rule 9 Capitalize the first word and all major words in titles of books, articles, movies, and television shows, and names of businesses.

        "The Cosby Show"   *Electronics Review*
        In Search of Excellence   Scientific Atlanta

——————— **NOTE** ———————

Titles of books are either in italics (if possible) or underlined.
Titles of articles from newspapers, magazines, movies, and shows are in quotation marks.

Rule 10 Capitalize historic events, famous places, holidays, and organizations.

        Tuskegee Institute   Labor Day
        World Trade Center   Space Age
        United States Senate   Mississippi River

Rule 11 Capitalize brand names but not product names.

        Panasonic turntable   Pontiac sports car
        Epson printer   IBM typewriter

Capital Letters

NOTE

Many companies and products on the market have unusual spellings of common words. Be careful to spell them as they are used by companies.

Compaq Portable II   LANLink
B&K-Precision   Printonix

Exercise 3 Change the incorrect lowercase letters to capital letters.

1. jim watches "hill street blues" every week.
2. larry always did well in math until he took beginning calculus.
3. our family celebrated independence day by visiting the smithsonian museum.
4. after reading several issues of *hands-on electronics,* i ordered a subscription.
5. nancy and tom were inducted into alpha beta tau last term.
6. all the teenagers in london were wearing levi jeans.
7. the english teacher told us to read a *tale of two cities.*
8. i looked through the heathkit catalog to find a project for my digital electronics class.
9. the lion's club sponsored a christmas party at the northside children's hospital.
10. he gave his wife a siamese kitten, and she gave him an airedale puppy.

APPENDIX 1

COMMON SYMBOLS AND ABBREVIATIONS

ELECTRONICS UNIT SYMBOLS
(from Berlin: *The Illustrated Electronics Dictionary*)

Unit	Symbol	Unit	Symbol
ampere	A	joule	J
ampere-hour	Ah	kelvin	K
ampere-turn	At	lambert	L
baud	Bd	lumen	lm
bel	B	lux	lx
coulomb	C	maxwell	Mx
decibel	dB	meter	m
degree (angle)	°	mho	mho
degree (temperature)		oersted	Oe
Celsius	°C	ohm	Ω
Fahrenheit	°F	radian	rad
kelvin	K	second	s
Electronvolt	eV	siemens	S
farad	F	tesla	t
gauss	G	var	var
gilbert	Gb	voltampere	VA
henry	H	watt	W
hertz	Hz	watt-hour	Wh
horsepower	hp	weber	Wb
hour	h		

Common Symbols and Abbreviations

METRIC SYMBOLS

Prefix	Symbol	Multiplier	Prefix	Symbol	Multiplier
exa-	E	10^{18}	centi-	c	10^{-2}
peta-	P	10^{15}	milli-	m	10^{-3}
tera-	T	10^{12}	micro-	μ	10^{-6}
giga-	G	10^{9}	nano-	n	10^{-9}
mega-	M	10^{6}	pico-	p	10^{-12}
kilo-	k	10^{3}	femto-	f	10^{-15}
deci-	d	10^{-1}	atto-	a	10^{-18}

TITLE ABBREVIATIONS

Title	Abbreviation
Chief Executive Officer	CEO
Director	Dir.
Doctor	Dr.
Honorable (judge)	Hon.
Junior	Jr. (abbreviated only if following a name)
Man	Mr.
Manager	Mgr.
Married woman	Mrs.
President	Pres.
Professor	Prof.
Reverend	Rev.
Superintendent	Supt.
Unmarried woman	Miss (not an abbreviation)
Vice-President	VP
Woman	Ms.

STATE ABBREVIATIONS (U.S. Postal System)

State	Abbreviation	State	Abbreviation
Alabama	AL	Montana	MT
Alaska	AK	Nebraska	NE
Arizona	AZ	Nevada	NV
Arkansas	AR	New Hampshire	NH
California	CA	New Jersey	NJ
Colorado	CO	New Mexico	NM
Connecticut	CT	New York	NY
Delaware	DE	North Carolina	NC
District of Columbia	DC	North Dakota	ND
Florida	FL	Ohio	OH
Georgia	GA	Oklahoma	OK
Guam	GU	Oregon	OR
Hawaii	HI	Pennsylvania	PA
Idaho	ID	Puerto Rico	PR
Illinois	IL	Rhode Island	RI
Indiana	IN	South Carolina	SC
Iowa	IA	South Dakota	SD
Kansas	KS	Tennessee	TN
Kentucky	KY	Texas	TX
Louisiana	LA	Utah	UT
Maine	ME	Vermont	VT
Maryland	MD	Virginia	VA
Massachusetts	MA	Virgin Islands	VI
Michigan	MI	Washington	WA
Minnesota	MN	West Virginia	WV
Mississippi	MS	Wisconsin	WI
Missouri	MO	Wyoming	WY

ABBREVIATIONS OF DAYS AND MONTHS*

Sunday	Sun.	January	Jan.
Monday	Mon.	February	Feb.
Tuesday	Tues.	March	Mar.
Wednesday	Wed.	April	Apr.
Thursday	Thurs.	May	May (no abbrev.)
Friday	Fri.	June	June (no abbrev.)
Saturday	Sat.	July	July (no abbrev.)
		August	Aug.
		September	Sept.
		October	Oct.
		November	Nov.
		December	Dec.

*Do not abbreviate days or months in formal letters or reports, except in footnotes or bibliographies.

ROMAN NUMERALS

0	. . .	30	XXX
1	I	40	XL
2	II	50	L
3	III	60	LX
4	IV	70	LXX
5	V	80	LXXX
6	VI	90	XC
7	VII	100	C
8	VIII	200	CC
9	IX	300	CCC
10	X	400	CD
11	XI	500	D
12	XII	600	DC
13	XIII	700	DCC
14	XIV	800	DCCC
15	XV	900	CM
16	XVI	1000	M
17	XVII	2000	MM
18	XVIII		
19	XIX		
20	XX		

APPENDIX 2

TIPS FOR TYPING AND WORD PROCESSING

If you have never had a typing class, you may feel handicapped when you are expected to turn in typed reports, letters, résumés, and memos. Finding the letters on the keyboard will be the biggest problem until you practice. Be sure that you allow yourself extra time for slow typing.

But there are other things that you need to know when you begin typing. Some terms that will appear in this appendix need to be defined.

- *Keyboard:* the panel of keys with letters, numbers, and marks
- *Uppercase:* capital letters (ABCD)
- *Lowercase:* small letters (abcd)
- *Indent:* moving in 5 to 10 spaces at the beginning of a paragraph
- *Double space:* leaving an empty line between each printed line

USING THE TYPEWRITER

A typewriter has a variety of keys for letters, numbers, and punctuation marks. Some of these may be interchangeable. For instance, a lowercase "L" can be used as a number "1." If two items appear on one key, the upper symbol will appear when you hold down the shift key and press the symbol key.

Rule 1 Indent 5 to 10 spaces at the beginning of each new paragraph. This is especially important when you are double spacing. Be consistent in the number of spaces you indent.

Rule 2 Start each sentence with a capital letter and end each sentence with a punctuation mark (. ! ?).

Rule 3

- Place any punctuation mark next to the word that precedes it (no space).
- Put quotation marks immediately next to the first and last words of the quote, but outside other punctuation.
- Leave one space after commas and semicolons.
- Leave two spaces after a colon or a period, question mark, or exclamation mark at the end of sentences.
- Do not space after the first period of abbreviations such as a.m., N.Y., U.S., and M.D.
- Do not leave a space before or after a hyphen, such as "seven-year-old boy."
- Use two hyphens to indicate a dash (I enjoy summer--June through September--more than any other season).

Example: We were thrilled to have a paid, three-day vacation! Is the M. L. King holiday--the third Monday in January--celebrated everywhere in the United States?

Rule 4

- Plan a border for each page. This means deciding on top, bottom, right, and left margins.
- Six lines are approximately 1 inch.
- Ten spaces are approximately 1 inch.
- Each sheet of typing paper (or printer paper) is about 66 lines down and 80 spaces (columns) across.

Rule 5 Double-space all reports; single-space letters, memos, and résumés.

USING A COMPUTER AND WORD PROCESSOR

Using a computer to write makes the typing process easier and faster than using a regular typewriter. Although each computer and word processor (WP) may have unique features and procedures, there are some general rules to follow. Some additional words need to be defined.

- *Diskette:* also called disk or floppy disk, used to store commands for the computer and store your writing
- *Word processor:* a diskette, or software, that turns the computer into a typewriter
- *Monitor:* the screen or TV that acts as your "paper" while typing
- *Cursor:* the flashing mark on the monitor that shows where you are
- *Disk drive:* a component with a slot for inserting a disk
- *Return key:* also called the continue or enter key, used like a carriage return on a typewriter
- *Automatic word wrap:* the ability of the word processor to keep your writing within preset margins
- *File:* any report, memo, letter, or document that you write
- *Default:* preset number or command which can be changed if desired

Word processors are written for individual brands of computers, such as IBM, Apple, or Radio Shack. Usually, there is not much room left on the WP disk for storing your files. You may wish to purchase your own storage diskette. Make sure that the diskette you are buying will work on the type of computer you will be using. A new diskette is like a wax record without any grooves. You will have to

format your new diskette before it will store files. Get help—do not try to do this by yourself. Each computer has its own format. Remember that a disk formatted for an IBM, for instance, cannot be used on an Apple.

Treat any disk carefully—it is fragile and has a limited life span. Some things that will destroy disks quickly are heat (including direct sunlight), moisture, dust or hair, magnetic fields, or crushing (yes, disks can be crushed). Store your disk in the jacket in which it came. The jacket has a special dust-resistant coating. Do not put your disk inside books or purses. The best way to carry disks around is in a hard, plastic case, available in most software stores, or in a briefcase with separate, protective pockets.

Most word processors are **menu driven,** which means that you select letters or numbers from a menu, or list of functions, such as "type," "print," or "save." You will sometimes wish to select a **help screen**, which will provide extra information or directions. Some functions are performed only if you hold down the **control key** while pressing a letter. Some functions operate by using a **toggle**—pressing a key turns the function on, pressing the key again toggles it off, much like a light switch. The capital-letter lock is usually a toggle.

You will be able to set your own margins, or you can use the **default** margins. Remember that there are 80 spaces across standard printing paper. Actually, the computer stops at 79. When setting **left** and **right margins,** use numbers on the ruler line on the screen. For instance, if you want a ten-space margin on both sides, you will set the left margin at 10 and the right margin at 70.

The **top** and **bottom margins** are set differently. You will usually enter the number of empty lines you want at the top and bottom, such as a top margin of 3 (lines) and a bottom margin of 3.

Opening a file means that you have to name or label your file first. Some word processors will not require you to name your file until you SAVE it. Most computers allow up to eight letters or numbers in the name. Try to name your file something that will remind you of what is in the file, such as "lab1."

Once you have named a file, the name will appear in the **catalog** or **directory** of the disk. Often, if you give your new file the same name as an old file, the old file will be erased. You may want to examine your directory before you name a new file.

The **load** or **get** function allows you to call up an "old file," a file that is already stored on the disk. You may add to or change your old files at any time. It is also possible, in most cases, to merge two or more files.

The **save** function allows you to store your file on the disk. It is a good idea to save your file periodically as you are writing it. The computer memory is temporary. The disk memory is permanent. If the computer loses power for any reason, what you have written on the screen is lost unless you have saved it on the disk.

Some word processors will show you on the screen exactly what your printed page will look like (what you see is what you get), and others will not. You may have to make a trial print, and then edit it.

The **automatic word wrap** feature means that you do not have to press the return key as you type close to the right margin. The computer will automatically "wrap" the last word around to the next line. You only need to press the return key when you have to start on a new line, such as at the beginning of a new paragraph. If you want to double space, ask an expert when and how to tell the computer to double space—each WP is different. Sometimes you **set line spacing** before you start typing, and the computer will automatically skip two (or more) lines for you. Other times it is a printing command (the printer double spaces what you have single-spaced).

Editing (changing, adding, or deleting) is usually done using **cursor control keys,** arrows keys that move the cursor without changing any letters. Most word processors have **insert** and **delete** keys that allow you to make changes quickly.

Printing your file or document is the last step. You must turn on the printer, and make sure that the printing paper is aligned (the "perf" or tear marks are lined up with the carriage of the printer). Sometimes the computer will ask you questions before it starts printing. Ask for help if you do not understand the questions. Most questions will have a **default** answer (press the **return key** to accept this answer) and will have other possible answers that could replace the default answer.

All word processors are purchased with a **manual** or some type of printed reference book to answer specific questions. Some have workbooks or demonstration disks (demos) to give guided practice in using the language. Some manuals are poorly written, making them bewildering and complicated to read and use. Others are written so that even inexperienced computer operators can easily understand basic operating rules. These manuals have been written by effective technical writers—who once learned basic technical writing just as you are now doing.

APPENDIX 3

TECHNICAL REPORT FORMAT

Some instructors (supervisors, companies) require specific formats for technical reports. If not, follow these general guidelines.

1. Type the report, double-spaced, on white, 8½ by 11 inch paper.
2. Leave a 1-inch border all around the paper (right, left, top, and bottom margins).
3. Capitalize headings and either center them or place them at the right margin (be consistent).
4. Indent every paragraph at least five spaces (be consistent).
5. Number pages (if the body of the report is over three pages) in the bottom margin, either centered or in the right corner (be consistent). Start numbering from the introduction to the report. Pages before the introduction are marked with lowercase Roman numerals.

    Some manuals and documentation are numbered using a decimal or hyphenated system. The chapter number is noted first, followed by the page within the chapter. At each new chapter, the page numbering begins with 1.

EXAMPLES OF PAGE-NUMBERING SYSTEMS

	Sequential	*Decimals*	*Hyphens*
Chapter 1	1	1.1	1-1
	2	1.2	1-2
Chapter 2	3	2.1	2-1
	4	2.2	2-2

6. Staple the completed report in the upper left-hand corner. Do not include any blank sheets or use a plastic cover (unless required). If the report is longer than 10 pages, put it in a three-ring binder.

TECHNICAL REPORT FORMAT
The outline for the format is as follows:

I. PRELIMINARY SECTION
    1. Title page
    2. Letter of Transmittal or Preface
    3. Table of Contents and List of Figures
    4. Abstract

II. MAIN SECTION
    5. Introduction
    6. Body
    7. Conclusion and Summary
    8. Tables and Figures (if not included in body)

III. DOCUMENTATION
    8. Quotations (footnotes or endnotes, if needed)
    9. Bibliography

1. Title Page. The title page, used for reports of three pages or more, is the first page of the report. Although styles vary, the information on this page includes the exact title, the author and position, the person to whom the report is submitted, and the date of submission. Arrange the information atractively and legibly.

2. Letter of Transmittal or Preface (optional). This page may be in memo or letter form (called a letter of transmittal), addressed to a specific reader. This page may also be in paragraph form (called a preface) for general readers. Use one or the other—not both. Provide background information: the reason for the report, the title of the report, special features of the report, and acknowledgement of any special assistance in the research/writing process.

3. Table of Contents and List of Figures and Tables. Each of these pages lists the order, topic headings, and page numbers for easy location by the reader. Include a table of contents for all reports over three pages. Include a list of figures and graphs if five or more formal graphics are included in the report. Arrange the information in neat columns. Use double-spaced dots to connect titles with page numbers.

4. Abstract (optional). The purpose of the abstract is identical to that of a summary—to provide a brief overview of the contents. The placement is the only difference—an abstract is located before the report begins, a summary is placed at the end. Do not include both. The length of the abstract should not be more than 10 percent of the length of the report. It should include the thesis (main purpose) and all main ideas in the same order as they are presented in the report.

5. Introduction and Thesis Sentence. The introduction is the first paragraph of the main section of the report, and it is always labeled. It provides a general lead-in to the subject of the report. This is the logical place to define a term, state the focus of a topic, or provide a brief background or history of the topic (assuming "History" is not a main heading in the report). The last sentence of the introduction is called the *thesis sentence*. The thesis sentence simply tells the reader the outline of the report.

    The purpose of this report is to discuss the history, function, and
    types of resistors.

The thesis sentence should match the sequence of the ideas that follow.

6. Body. This is the longest part of the report. It is made up of several individual sections with headings and includes many paragraphs. The order or sequence of the sections will depend on your purpose and topic. Each section must begin with a new paragraph. Do not begin a section or any new paragraph with a pronoun.

Labeling each major section of your report will provide a clear direction for the reader. Some common labels that state the purpose and the scope of sections are listed below.

HISTORY	TYPES	OPERATION
APPLICATIONS	CONSTRUCTION	FEATURES
USES	DESCRIPTION	FUNCTION

To make the report easy to read, keep sentences under 20 words, and paragraphs under 10 lines of print (a suggestion, not a rule). To make the report interesting, vary your sentence length and style, using the methods you have practiced in sentence combining exercises. Finally, proofread and edit carefully for clarity, grammar, spelling, and word choice. The revision process is often the most important step in effective writing.

7. Conclusion and/or Summary. The ending of the report must be clearly labeled and include at least one paragraph. A *summary*, like the abstract, is a brief review of the main ideas—no new information is provided. A *conclusion*, however, after restating the thesis, may include a personal opinion, analysis, or recommendation based on the research of the topic. At times, both endings are appropriate. The summary should come first, followed by the conclusion. Label each separately.

8. Quotations. Many authors find it necessary to "borrow" facts, statements or figures from other authors. The careful and limited use of quotations is accepted as long as the writer does two things: place quotation marks around the exact words taken from another author (not necessary if the information is paraphrased); and cite the source of the quote (or paraphrased version) either by a footnote, endnote, or parenthetical note. Failing to do these things is considered *plagiarism,* passing someone else's work off as your own. It is a serious offense and a breach of ethics in technical writing.

If *footnotes* are used (the traditional form of citing a quotation), a superscript number is placed at the end of the quote. At the bottom of that page, write the superscript number and the footnote information for each quote used on the page. Number the citations consecutively throughout the report.

Although the exact entry form differs slightly depending on the source, generally start with the author's name (first, last), the title, publishing information, date, and page(s). If more than one location is given, list the closest to you. If more than one publication date is given, list the more current. Indent the first line of the entry, and outdent the second (third, etc.) line. *Remember to underline titles of books or magazines.*

Samples of footnotes for information taken from the following:

1. Book
    [1]Samuel L. Oppenheimer, *Fundamentals of Electric Circuits* (Englewood Cliffs, NJ: Prentice-Hall, Inc., 1984), 290.
2. Magazine article
    [2]Arthur Fisher, "Science Newsfront," *Popular Science* February 1986: 13–15.
3. Encyclopedia article
    [3]Christos Halkias, "Direct-coupled amplifier," *McGraw-Hill Encyclopedia of Science and Technology,* 1982 ed.
4. Manual
    [4]*8010A/8012A Digital Multimeter Instruction Manual,* John Fluke Manufacturing Company, Inc., 1985.

If *endnotes* are used, number the quotes consecutively throughout the report and cite the quotes at the end of the report on a page labeled "Endnotes," "Notes," or "References." Use the same format as footnote entries. If endnotes cite all references used for the report, a bibliography is usually unnecessary. Check with your instructor.

The newest format for citations, recently adopted by the Modern Language Association and the Society for Technical Communication, is the *parenthetical note*. This form simply cites the author's last name (if none, a key word from the title) and a page number in parentheses outside the final quotation mark, but inside the final punctuation of the sentence. Its efficiency makes it the best choice.

Sample of parenthetical note:

> "Inductance is that property of a circuit that opposes changes in a circuit" (Oppenheimer, 290).
>
> Extra precautions must be taken not to exceed the 10A maximum current capacity of the 8010A (*Multimeters*, 2.4).

This reference says that the quotation about inductance is taken from page 290 in the book by Oppenheimer listed in the bibliography of this report. Then list a complete entry in the bibliography.

The second entry is an indirect (paraphrased) quote. No quotation marks are needed, but the citation is included. Again, the full citation should be listed in the bibliography.

9. *Bibliography*. The bibliography is normally the last page of a report. Its purpose is to list all the references you used to research your report, whether you quoted from them or not. References can include books, magazines, encyclopedias, newspapers, manuals, documents, lecture notes, or personal interviews. Dictionaries are normally not considered references and are not listed in the bibliography. Any source from which a quote was used in the report must be listed in the bibliography.

If the entry is longer than one line of print, single-space and indent the second (third, etc.) line of the entry. Double or triple space between entries.

Entries in the bibliography contain the same information as footnotes, but in a slightly different format. They are arranged in alphabetical order by the first word of the entry, usually the author's last name. The first line of the entry is outdented and each subsequent line is indented. The footnote entries listed above would be arranged in a bibliography as follows.

Samples of the entries in a bibliography:

1. *Book*
    Oppenheimer, Samuel L. *Fundamentals of Electronic Circuits*. Englewood Cliffs, NJ: Prentice-Hall, Inc., 1984.
2. *Magazine article*
    Fisher, Arthur. "Science Newsfront." *Popular Science*. February 1986: 13–15.
3. *Encyclopedia*
    Halkias, Cristos. "Direct-coupled amplifier." *McGraw-Hill Encyclopedia of Science and Technology*, 1982 ed.
4. *Manual*
    8010A/8012A Digital Multimeter Instruction Manual. John Fluke Manufacturing Company, Inc., 1985.

> *Note:* Alphabetize numbers by the spelling of the first word (8010A—Eight thousand . . .)

Consult your instructor, manager, or a current style guide, such as the *MLA Handbook for Writers of Research Papers* (by Joseph Gibaldi, New York: Modern Language Association, 1984) about specific formatting questions.

The following technical report shows many of these writing conventions.

THE GREENHOUSE EFFECT

by

JERRY LOPEZ

Ms. A. Rutherfoord

CS-31

January 12, 1987

Decatur, Georgia
January 12, 1987

Ms. Andrea Rutherfoord
Department of General Education
DeVry Institute of Technology
Atlanta, Georgia 30030

Dear Ms. Rutherfoord:

I am submitting this report to meet the requirements of Communication Skills (CS-31). In this report, I have examined the greenhouse effect, with emphasis on the predictions made by climatologists concerning the future impact of current changes in the atmosphere.

The background of the phenomenon has been obtained from encyclopedias, but the primary source of significant data is from the *Global 2000 Report to the President*. Written in 1982, this document addresses the concerns of former President Jimmy Carter. It was his commitment to our quality of life and the preservation of life that spurred scientists to examine the effects of various changes in our environment.

I believe that this report will be useful in stressing the fragile balance which allows life on earth to continue. Awareness is the first step in correcting and elmininating life-threatening practices.

An abstract follows which explains my approach in this report.

Sincerely,

Jerry Lopez

Jerry Lopez

CONTENTS

Letter of Transmittal ii

Abstract iv

Introduction 1

Cause-and-Effect Relationship 1

Components of the Greenhouse Effect 1

The Future 2

Conclusion 2

Bibliography 3

ABSTRACT

    The greenhouse effect is a behavior in which the atmosphere acts as a greenhouse glass roof to control and protect the earth's temperature from the sun's lightwaves. New particles which are introduced into the atmosphere threaten to destroy the delicate balance necessary for the greenhouse effect to sustain plant and animal life on earth.

INTRODUCTION

In recent years, there has been a lot of publicity over the greenhouse effect. This publicity may lead some to believe that the greenhouse effect is a product of man, but that is simply not true. The greenhouse effect is what makes the earth's atmosphere temperate enough to support human life. In a television broadcast of "Carl Sagan's Cosmos," the narrator discussed the possibility that changes in the greenhouse effect caused the extinction of the dinosaurs. In order to gain a circumspective view of this life-sustaining phenomenon, it is necessary to examine the cause-and-effect relationship that makes up the greenhouse effect, the components of the relationship, and predictions that have been drawn by scientists regarding the greenhouse effect.

CAUSE AND EFFECT

The greenhouse effect is due to the ability of molecular elements, suspended in the earth's atmosphere, to block ultraviolet wavelengths and trap infrared wavelengths. This causes the average global temperature to stay between the freezing point and boiling point of water.

The sun emits energy in many different wavelengths. Some of this energy (ultraviolet wavelengths) is not good for carbon-based life on earth. The atmosphere blocks these harmful emissions. It is our atmosphere that lets sunlight (infrared wavelengths) through to the surface of the earth which creates heat. The heat, however, cannot pass out of the atmosphere easily. As a result, the earth is warmed and life is sustained through a balance of sunlight and atmospheric particles (*World Book*, p. 783).

Thus the atmosphere acts like the glass roof of a greenhouse, which creates a climate necessary to sustain the life within the greenhouse. The introduction of particles not normally found in the atmosphere can potentially cause an imbalance within the greenhouse effect, thus endangering plant and animal life on earth.

COMPONENTS OF THE GREENHOUSE EFFECT

The components of the greenhouse effect are of major importance to man. The particles known as ozone are found in the upper portion of the atmosphere. These particles tend to block the ultraviolet light emitted by the sun. If ultraviolet light were to reach the earth's surface in abundance, it would cause most carbon-based life to die. A small increase in ultraviolet light is suspected to be a cause in the increased number of cases of skin cancer in the human population.

Water and carbon dioxide are the major elements in the warming effect of the earth. Changes in these levels could possibly cause a loop reaction. For example, an increase in carbon dioxide could cause a warming trend which would then cause an increase in the melting of the polar ice caps. Water vapor has a warming effect, thus causing a further warming.

THE FUTURE

Dinosaurs became extinct long before human beings first appeared, yet modern man's habits threaten the greenhouse balance which supports life today. People cannot control nature, but they can hopefully gain the wisdom needed to at least avoid causing the complete annihilation of life on earth.

In 1977, then-President Jimmy Carter commissioned a group of scientists and officials to study the probable changes in the world's population, natural resources, and environment through the end of the century. They produced *The Global 2000 Report to the President* which summarized the responses of 24 major climatologists from seven countries. One scenario resulting from the survey predicted a large global warming by the end of the century, which would bring with it a greater likelihood of continental drought in the United States.

The human race uses huge amounts of fossil fuels today. One by-product of burning fossil fuels is carbon dioxide, one of the elements which interacts in the greenhouse effect. Carbon dioxide levels in the atmosphere have increased by 5% in the past 20 years (*Global 2000*). "Since the total amount of carbon in the earth-atmosphere system is constant, the carbon being added to the atmosphere pool must come from a nonatmospheric carbon pool somewhere within the system" (*Global 2000*, p. 261).

Some scientists suspect that the additional carbon is the by-product of burning fossil fuels. They feel that this increase threatens to tip the delicate balance in the greenhouse effect necessary to sustain life on earth.

It is impossible at this time for scientists to predict what will happen to our climate in the future, but, as the *Global 2000* report goes on to say, "Many experts nevertheless feel that changes on a scale likely to affect the environment and the economy of large regions of the world are not only possible but probable in the next 25 to 50 years" (p. 257).

CONCLUSION

With this new information, we can conclude that the human race can create its own fate. Can we gain the wisdom and take the necessary steps to ensure our survival, or will we ignore our adverse effects on nature's balance and cause our own extinction? Education has, in the past, led the human race to achieve and triumph over conditions beset upon it by nature. Once again, education will be the answer.

BIBLIOGRAPHY

Council on Environmental Quality and the Department of State. *The Global 2000 Report to the President.* New York: Penguin Books, 1982.

"Greenhouse Effect," *Academic American Encyclopedia,* Vol. 9. Danbury, CT: Grolier Inc., 1986.

"Sun," *The World Book Encyclopedia,* Vol. 18. Chicago: World Book-Childcraft International, 1981.

APPENDIX 4
SPELLING AND MISUSED WORDS

aberrant, aberration
abscissa
absorb, absorption
acceptor
accessible
accommodate
accuracy
achromatic
acknowledgement
across
admissible
aerosol
afterward
age, aged, aging
air-cooled
air-dried
align, alignment
a lot (not alot)
alpha
ammeter
ampere
ancillary
anonymous
aperture
arc, arced, arcing
asterisk
asymmetric
audible
audio-frequency
auxiliary
axes (pl. of axis)
bandwidth
bargain
battery
beneficial
benefited
beta
beveled
biased
binary
Bohr-Sommerfeld atom
breadboard
bypass
by-product
cannot
cassette
catalog
channeled
characteristic
chassis (s. and pl.)
circuit
coherent
cohesive
collapsible
collectible
combustible
compatible
comprehensible
comprise
condensable or condensible
conductor
cooperate
coordinate
corollary
coulomb
co-worker
crises (pl. of crisis)
criteria (pl. of criterion)
crystal
curvilinear
cylinder
data (pl. of datum)
defensible
diagnosis
diagrammed
dielectric
diesel
diffusible
dimensional
disassemble
disk, diskette
eccentric
efficient
electrode

electrolyte
electronic, electronics
eligible
embarrass
emitter
engineer
exhaustible
expandable
fascinate
feasible
flux
Fourier
frequency
fulfill or fulfil
fundamental
fusion
gauge
geometric
germanium
guaranteed
gyrator
harmonic
height
henry, henries
hertz (s. and pl.)
horizontal
hybrid
hysteresis
illogical
imaginary
immediately
impedance
incident
indispensable
inductance
insertion
instantaneous
instrument
insulation
integrated
intensity
intermittent
interruption
inversely
isolation
isotropic
joule
judgment or judgement
junction
kelvin
Kirchhoff
knowledge
leakage

leisurely
length
license
linear
liquid
logarithm
logarithmic
luxury
magnetic
majority
maneuver
match
meant
mercury
microprocessor
minority
modulation
multiplex
multiplier
mutual
necessary
neutralize
neutron
nickel
nuclear
occasional
occur
occurrence
ohmmeter
omission
omitted
opportunity
optimistic
oscillate
oscilloscope
parallel
paraphrase
particularly
pastime
peculiar
permanent
permeability
permeance
permissible
pleasant
polarized
possibly
potentiometer
privilege
procedure
professor
prognosis
programmable

programmed or programed
questionnaire
quiescent
radiation
received
reciprocal
recommend
reference
rehearsal
repetition
resistance
resistivity
resistor
resonance
resonant
rotary
rough
satellite
saturation
scarcity
schedule
schematic
secretary
sensitivity
separate
siemen
silicon
sincerely
socket
solder
spectrum
squelch
statistics
straight
strategy
summarized
susceptible
synchronous
technical
technician
technique
tendency
therefore
thermionic
thermostat
transducer
transformer
transient
transistor
transmitter
transponder
tubular
tweeter

Spelling and Misused Words

ultrasonic	velocity	wire-wound
ultraviolet	vertical	woofer
universal	vibration	X-axis
until	video	X-band
useful	visibility	X-ray
usually	voltage	Y-axis
vacuum	volume	Y-junction
variable	wattmeter	Z-axis
vector	wavelength	Zener

MISUSED OR CONFUSED WORDS

a, an, and	decent, descent, dissent	official, officious
accept, except	desert, dessert	past, passed
advise, advice	device, devise	perfect, prefect
affect, effect	dyeing, dying	perpetrate, perpetuate
alternate, alternative	elapse, lapse, relapse	personal, personnel
analog, analogous	elicit, illicit	peruse, pursue
analysis, analyze	eminent, imminent	picture, pitcher
anonymous, unanimous	ensure, insure	precede, proceed
access, excess	envelop, envelope	quiet, quit, quite
all ready, already	expect, suspect	recent, resent
all together, altogether	fair, fare	sign, sine
angel, angle	farther, further	stationary, stationery
born, borne	from, than	suppose, supposed to
brake, break	ladder, later, latter	then, than
caliber, caliper	least, lest	they're, there, their
causal, casual	lose, loose, loss	thorough, though,
command, commend	moral, morale	thought, tough
complement, compliment	morality, mortality	to, too, two
conscience, conscious,	knew, new	threw, through
conscientious	know, no	use, used to
continual, continuous	of, off	wear, were, we're

APPENDIX 5

IRREGULAR VERBS

Present	Past	Past Participle (Used with Have, Has, or Had)
be, is, am, are	was, were	been
beat	beat	beaten
become	became	become
begin	began	begun
bend	bent	bent
bid (price)	bid	bid
bid (command)	bade	bidden
bind	bound	bound
bite	bit	bitten
bleed	bled	bled
blow	blew	blown
break	broke	broken
bring	brought	brought
build	built	built
burst (not bust)	burst	burst
buy	bought	bought
cast	cast	cast
catch	caught	caught
choose	chose	chosen
cling	clung	clung
come	came	come
cost	cost	cost
creep	crept	crept
cut	cut	cut
deal	dealt	dealt
dig	dug	dug
dive	dove, dived	dived
do	did	done
drag	dragged (not drug)	dragged
draw	drew	drawn
drink	drank	drunk

326

Present	Past	Past Participle (Used with Have, Has, or Had)
drive	drove	driven
eat	ate	eaten
fall	fell	fallen
feed	fed	fed
feel	felt	felt
fight	fought	fought
find	found	found
fly	flew	flown
forget	forgot	forgotten, forgot
freeze	froze	frozen
get	got	got, gotten
give	gave	given
go	went	gone
grow	grew	grown
hang (execute)	hanged	hanged
hang (suspend)	hung	hung
have	had	had
hear	heard	heard
hide	hid	hidden
hit	hit	hit
hold	held	held
hurt	hurt	hurt
keep	kept	kept
know	knew	known
lay (place)	laid	laid
lead	led	led
leave	left	left
lend	lent	lent
let	let	let
lie (falsehood)	lied	lied
lie (recline)	lay	lain
lose	lost	lost
mean	meant	meant
prove	proved	proved, proven
read	read	read
rise	rose	risen
run	ran	run
say	said	said
see	saw	seen
sell	sold	sold
set	set	set
shake	shook	shaken
shine (glow)	shone	shone
shine (polish)	shined	shined
shoot	shot	shot
show	showed	shown, showed
shrink	shrank	shrunk
sit	sat	sat
speak	spoke	spoken
spin	spun	spun
spring	sprang, sprung	sprung
swear	swore	sworn
take	took	taken
teach	taught	taught
tell	told	told
think	thought	thought
throw	threw	thrown
troubleshoot	-----	-----
wear	wore	worn
win	won	won
write	wrote	written

INDEX

A

A/AN/AND, 18
Abbreviations, 299–300
    list of common, 305–7
Abstract, 32
ACCEPT/EXCEPT, 109–10
Acronyms, 300
    plurals, 14
Adjectives, 248–52
Adverbs, 252
ADVICE/ADVISE, 139–40
Agreement, 229
    with indefinite pronouns, 243–44
    pronoun reference, 245–47
    subject/verb, 229–33, 236–37
Analogies, 44–46
    review exercise, 95
ANCE/ENCE endings, spelling tips, 107–8
Apostrophes, 284–87
    review exercise, 93

B

Brackets, 296

C

Capital letters, 301–4
Cause-and-effect, 82–83
Classification, 9–10
Clauses, 210, 215, 220
    independent, 210, 215, 220
    subordinate (dependent), 220–27
Clichés, 44
Colon in lists, 10
    other uses, 293–94
Commas, 58, 275–83
    with adjective pairs, 276–77
    in compound and complex
        sentences, 217–18, 222–23, 281–82
    review chart, 58
    with interruptors, 277–79
    in introductory information, 280–81
    miscellaneous uses, 282–83
    in phrases, 278
    with quotes, 281
    in series, 275–76
Comparison and contrast, 65–67
Complex sentences, 220, 228
    punctuation of, 222–23
COMPLIMENT/COMPLEMENT, 189
Compound-complex sentences, 227–28

329

Compound sentences, 215–19
  punctuation of, 217–19
Contractions, 284–85
Crossword puzzles, 56–57, 94–95, 148–49, 196–97

D

Dash, 296
Definitions, 4–6
  formal, 5
  informal, 5
Description, 63–65
Descriptive report, 102–4
  example of, 104–6
Directions, written, 135–36
  example of, 136
Dropping the final E spelling rules, 84–85
Double letter problems, spelling tips, 125–26
Double negatives, 252–54
Doubling the final consonant rules, 47

E

EFFECT/AFFECT, 86–87
EI/IE spelling rules, 34
Ellipses, 297

F

Forbidden words and phrases, 180–81
Formal lab report, 75–77
  example of, 77–81
Forms, 156–60
Fragments, 210–14

G

GRAD/GRESS roots, 188–89
GRAPH/GRAM roots, 126–27
Graphics, 115–23
  block diagram, 118
  cutaway drawing, 118
  graphs, line and bar, 120
  line drawing, 116
  photograph, 116
  pictorial diagram, 117
  pie chart, 121
  process drawing, 118
  tables, 122
Greek number roots, 17

H

Hyphen, 297–98
Hyphenated words, 15

I

IBLE/ABLE endings, spelling tips, 186–88
Inferred you, 205
INTRO/INTRA/INTER roots, 70–71
ISE/IZE/YZE endings, spelling tips, 137–38

L

Letters, business, 175–86
  examples of blocked, 176, 183
  examples of semiblocked, 177, 185
  negative, 184–86
  positive, 182–84
Lists in sentences, 10
LOSE/LOST/LOSS/LOOSE/LOOSEN, 128–29
LY/LLY endings, spelling rules, 25–26

M

Memos, 156, 160–62
  example of, 161
MICRO/MACRO roots, 139
Misused words, list of, 325
  (see also Word Watch sections in chapters 1–15)
Modifiers, 248–55
  misplaced, 254–55
MONO/BI/SEMI/POLY roots, 85–86

N

Negative prefixes, 26–27
Notetaking, 11–13
Nouns, 201
  pronouns, 201
Numbers:
  Greek roots, 17

Latin roots, 16
rules for spelling in writing, 67–69

O

"OUGH" words, 71–72
Outlining, 9

P

Parallelism, 256–60
Parentheses, 295
Participles, 204
   present and past, 204
PAST/PASSED, 179–80
Plagiarism, 288, 291
Plurals, 13–16
   acronyms, 14
   F plurals, 14
   hyphenated words, 15
   irregular, 13, 15
   nouns, 13
   -SIS to -SES, 15
   Y plurals, 13
Possessives, 285–87
Prepositions, 235–37
Process report, 135–36
   examples of, 136
Pronoun reference, 245–47
Pronouns, 240–44
   compound, 241–42
   indefinite, 243–44
   object, 241
   possessive, 242
   reflexive, 242–43
   subject, 240–41
PROTO/TRANS/NEO roots, 108–9

Q

Quotation marks, 288
   direct quotes, 288–90
   indirect quotes, 290–91
   other uses, 292

R

Redundancy, 42, 181
RETRO/CIRCUM roots, 70–71

S

Scientific method, 75
SEDE/CEED/CEDE roots, 167–70
Semi-colon, 294
   in a series, 10
   in compound sentences, 218–19
Sentence combining, 207–8
   exercises, 90, 208, 233, 237
Sexism in writing, 267–69
Shifts:
   tone, 264–66
   verb and pronoun, 261–62
   voice, 262–64
Signal words, 220–26
   (see *Subordinate Conjunctions*)
Slang, 41–43
Slash, 297
SPEC/SON roots, 35
Spelling words, list of, 323–35
   (see *also Spelling sections in chapters*)
Standard English, 42
STATIONARY/STATIONERY, 189–90
Subjects, 201–3
Subject/verb agreement, 229–33
   with indefinite pronouns, 243–44
   with prepositional phrases, 236–37
Subordinate conjunctions, 220–26
SUB/SUPER roots, 48–49
Summary, written, 32–33
Syllables, 47

T

Technical report format, 312–15
   example of, 316–22
TELE/PHONO/PHOTO roots, 126–27
THEY'RE/THEIR/THERE, 35–36
Tone shifts, 264
Topic sentences, 23
TO/TOO/TWO, 19
Transition words, 270–72

U

USED/SUPPOSED, 49–50

V

Verb errors, 42
Verbs, 203–5

Verbs (*cont.*)
   helping and linking, 203
   inferred you subject, 205
   irregular, 204
   list of common irregular, 326–27
   participles, 204
   regular, 204
Verb shifts, 261

Voice:
   active and passive, 8
   shifts, 262–64

W

WEAR/WERE/WE'RE/WHERE, 28
Word processing tips, 308–11